The functional response to prey density in an acarine system

Simulation Monographs

Simulation Monographs is a series on
computer simulation in agriculture and
its supporting sciences

The author works at the Netherlands Institute for Sea Research,
Texel, the Netherlands

The functional response to prey density in an acarine system

H.G.Fransz

Wageningen
Centre for Agricultural Publishing and Documentation
1974

Books already published in the series:

C. T. de Wit and H. van Keulen
Simulation of transport processes in soils
1972, 108 pages, ISBN 90 220 0417 1, price Dfl. 15.00

J. Beek and M. J. Frissel
Simulation of nitrogen behaviour in soils
1973, 76 pages, ISBN 90 220 0440 6, price Dfl. 12.50

C. T. de Wit and J. Goudriaan
Simulation of ecological processes
1974, 167 pages, ISBN 90 220 0496 1, price Dfl. 22.50

ISBN 90 220 0509 7

The author graduated on 3 May 1974 as Doctor in de Landbouwwetenschappen at the Agricultural University, Wageningen, the Netherlands, on a thesis with the same title and contents.

Cover design: Pudoc, Wageningen

Printed in Belgium

Contents

1   Introduction   1

1.1   The problem   1
1.2   An approach by simulation   3

2   The predator-prey system   5

2.1. A general description   5
2.2   Components of prey behaviour   6
2.3   Components of predator behaviour   7
2.4   A preconceptual model of the predator-prey system   7
2.4.1 The encountering rate   8
2.4.2 The success ratio   9
2.4.3 The length of the handling periods   9
2.4.4 The coincidence in space of the predator and the prey   9
2.4.5 The locomotion activity of the predator and the prey   10
2.4.6 The locomotion velocity of the predator and the prey   11
2.4.7 The state variables of the system   11

3   The analysis of the structural relationships   15

3.1   Materials used for observing and recording the predation
      process   16
3.2   Standardization of the predator   16
3.3   The predator's ingestion rate   17
3.4   The predator's digestion rate   22
3.5   The relative density of the webbing cover produced by the
      prey   24
3.6   The measurement of the locomotion velocity   27
3.7   Continuous observation of the predation process and the
      analysis of the records of events observed   28
3.8   A comparison of standard predators and ovipositing pre-
      dators   32
3.9   Polyfactor analysis of the multiple relationships between the
      system elements   34

3.10 A conceptual model of the predator-prey system 45

4   Simulation   46
4.1 Updating   46
4.2 Stochastic versus deterministic simulation models   47
4.3 Principles of compound simulation applied to classes of
     individuals   54
4.4 Simulation models   57
4.4.1 Stochastic simulation of the predation on prey eggs   57
4.4.2 Deterministic simulation of the predation on prey eggs   68
4.4.3 Compound simulation of the predation process   70
4.5 Simulated responses to prey density   79

5   Discussion   81

5.1 The functional response to prey density   81
5.1.1 The predation of eggs   84
5.1.2 The predation of males   86
5.1.3 The predation of eggs and males together   88
5.2 The influence of some factors on the functional response   91
5.2.1 Adaptation of predators to a given prey density   91
5.2.2 Hunger of the predator   93
5.2.3 Inhibition by prey   96
5.2.4 The density of the webbing cover   97
5.2.5 Subsidiary effects of the prey male density   101
5.2.6 Consequences of polyphagy   101
5.2.7 The interval of prey replenishment   108
5.3 The role of chance in the predation process   109
5.4 Implications of the functional response derived   110
5.4.1 The numerical response to prey density   110
5.4.2 Regulation of spider mite populations   111
5.5 Application of the results in population models   115

Summary   117

Appendices   122

Acknowledgments   138

References   139

1 Introduction

1.1 The problem

The role played by predacious mites in the natural control of phytophagous mites has been emphasized by many authors. In a review
on the ecology of tetranychid mites and their natural enemies, Huffaker
et al. (1970) concluded that populations of predacious mites may be
utilized successfully against harmful mites. However, the application
of control measures, such as the adjustment of chemical control practices or the release of suitable mite species, is impeded by lack of
knowledge of the conditions for natural control by predacious mites.
An essential condition for the control by predators is the existence of
a regulating mechanism, i.e. the mortality of the prey population due
to predation must exceed its reproduction, when the prey density
increases beyond a certain value. Hence, regulation will occur when
the predators effect an increasing mortality rate as the prey density
rises. The prey mortality due to predation is the number of prey killed
per unit of time and leaf area, which equals the product of the predator
density and the number of prey killed per unit of time per predator.
Both factors of the product may depend on prey density. Their relationships with prey density have been termed the numerical response
and the functional response of the predator population to prey density
(Solomon, 1949).
The functional response contributes to regulation, when the proportion of prey killed increases with prey density at a constant predator
density. It also may result in an increased reproduction rate of the
predator, leading to a numerical response. Therefore the functional
response is an important phenomenon. Three fundamental types of
functional response curves are described by Holling (1959):
Type 1: a linear rise to a plateau (Fig. 1A).
Type 2: a negatively accelerated rise to a plateau (fig. 1B).
Type 3: an S-shaped rise to a plateau (Fig. 1C).
Only a Type 3 response can induce regulation directly.
The functional response of predacious mites to the density of phyto-
phagous mites has been studied experimentally in laboratory systems

1

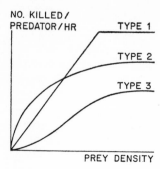

NO. KILLED/
PREDATOR/HR

TYPE 1

TYPE 2

TYPE 3

PREY DENSITY

Fig. 1 | Fundamental types of functional
response curves distinguished by Holling.

by Chant (1961b), Kuchlein (in prep.), Mori (1969), Mori & Chant
(1966) and Sandness & McMurtry (1970). All observers used a single
adult female as a predator on isolated areas with different prey densities.
The curve obtained by Chant for *Typhlodromus occidentalis* preying
on protonymphs of *Tetranychus telarius* is intermediate between
Holling's Type 1 and Type 2 responses. Mori and Mori & Chant,
observing *Amblyseius longispinosus* and *Phytoseiulus persimilis* preying
on deutonymphs and adult females of *Tetranychus urticae*, found
domed curves. Such curves resemble a Type 2 response, but have a
declining slope at high prey density. The curves obtained by Sandness &
McMurtry for *Amblyseius largoensis, A. concordis* and *Typhlodromus
floridanus* preying on adult females of *Oligonychus punicae* have a
curvilinear rise to a plateau at intermediate prey densities, and a rise
to a second plateau at high prey densities. Kuchlein's observations
on *Typhlodromus occidentalis* with eggs and males of *Tetranychus
urticae* as prey reveal for eggs a functional response resembling the
one found by Chant, and for males those found by Sandness &
McMurtry. It seems clear that there is little uniformity in the functional
response in mite systems, and there are often deviations from the fundamental types distinguished by Holling.
One way to investigate the circumstances, under which predatory mites
are able to keep phytophagous mites on a low, economically acceptable
density level, is to study the response of the predators to changes of
prey density. This way has led to some types of functional response
curves, which by themselves do not give much information about
regulation in natural mite systems. The next step in this approach is
a causal analysis of the functional response to prey density. It has to

2

be made clear, which properties of the predator and the prey determine the shape of the curves. Then it may be possible to conclude, under which conditions these properties will induce regulation. This book presents an analysis of the predation process in the laboratory system studied by Kuchlein.

1.2 An approach by simulation

A general description of a predation process is given by Holling (1959a, 1961, 1966). During the time the prey is exposed to the predator, the latter alternately searches for and handles prey, although it may pause for resting or digestion. The duration of the activities of the predator is determined by components of the predator and prey behaviour, which in turn depend on variables like the hunger of the predator and the actual prey density. Such variables, named state variables, usually change with time and they define the state of the system at every instant. They have to be distinguished, because they represent the more substantial elements of a system. Their change is characterized by a flow of matter, energy or more abstract quantities; this flow can be defined by a differential equation. The periods of different activities of the predator, the components of behaviour, and the state variables are the elements of a predator-prey system. Numerous interrelationships between the elements determine the number of handling periods (killings) per time unit.

Holling (1966), analysing the functional response of the praying mantid *Hierodula crassa* to the density of houseflies, determined the relationships of relevant elements by an experimental analysis, and constructed a comprehensive mathematical model of this predator-prey system. With the aid of the model a computer could be programmed to simulate the predation process at different prey densities by computing series of alternate searching and handling periods. So the functional response can be related to invariable properties of the predator and the prey. The shape of the functional response curves observed in mites suggests that the mite system is at least as complex as the praying mantid-fly system. The mite system studied in this paper is described in Chapter 2, and it is indicated that computer simulation is a necessary tool in the analysis of the predation process. In that chapter, significant state variables and components of predator and prey behaviour are selected to be incorporated into a simulation model. Chapter 3 continues

3

Holling's approach with an experimental analysis of the structure of the system.

The mathematical model, outlined by Holling (1966), is proposed to be a general model, which can be adapted to any predation process. It computes searching periods, which are a function of state variables such as the hunger level of the predator. The values of the state variables at the beginning of the periods can be computed, but of course the variables change during the searching periods, and so do the searching periods themselves consequently. The model essentially requires the solution of a complicated differential equation, which determines the rate of change of the searching periods. This makes Holling's method mathematically cumbersome and even impracticable for more complex systems. Its general applicability is questionable.

Other, more flexible methods of computer simulation progress in time by small intervals, during which the state variables are considered to be constant. Only the rates of change of the state variables, which are simple functions of these variables, have to be considered. Numerical integration using these rates updates the state variables after each time-interval by standard procedures. Computer simulation languages using numerical integration have been developed for continuous processes. Their flexibility, however, allows their application in the simulation of discontinuous predation processes. Chapter 4 describes the application of Continuous System Modeling Program (CSMP), a simulation language developed by IBM (1968), in the construction of models for simulation of the predation process outlined in Chapters 2 and 3.

Attention is paid to the stochastic character of the process with a single predator, in which discrete events have a certain probability to occur in a certain time-period. One model simulates the predation on eggs by generating random events in each time-interval. Because this model has a variable output, it has to be rerun repeatedly to find the expectation values. To evaluate the role of chance, the stochastic model is compared with a deterministic one. A third model avoids the problems inherent in the others by computing expectation values of the output in one run with a proper account of the role of chance. This model is extended to include the presence of prey males as well as prey eggs. Finally, Chapter 5 discusses the results of simulation and their implications for the regulation of prey density. It also gives an explanation of the functional response curves obtained by Kuchlein.

4

2 The predator-prey system

2.1 A general description

The system to be modelled is the one used by Kuchlein (in prep.) to study the functional response of *Typhlodromus occidentalis* Nesbitt (= *Metaseiulus occidentalis* (Nesbitt)) (Acarina: Phytoseiidae) to the density of *Tetranychus urticae* Koch (Acarina: Tetranychidae). Very briefly this system can be described as follows:

Disks with an area of 5 cm² are punched from leaves of the lima bean, a variety of *Phaseolus vulgaris*, and are placed upside down on wet cotton-wool in Petri dishes. A number of adult males and/or fresh (translucent) eggs of *T. urticae* are transferred to the disks from rearing leaves by means of a brush. One three-to-ten-day-old adult, mated female of *T. occidentalis* is released per disk. The Petri dishes are stored in a climatic cabinet at 27 °C and 70% r.h. After about twenty hours the prey killed is replenished, and subsequently once every thirty minutes, for a period of six hours, the number of prey killed is counted under a binocular microscope and then replenished.

The globular prey eggs have a diameter of 0.127 mm and stick to the leaf surface. The males are somewhat conical in shape, caudally pointed, measuring 0.26 mm in length. The predator has originally been introduced in the Netherlands from north-east America (Bravenboer, 1959). It was collected by Kuchlein from greenhouses at Naaldwijk, the Netherlands, and kept in stock on bean plants with *T. urticae* as food. Adult females are ovular in shape and about 0.35 mm long. A picture of the animals used is given in Plate I.

T. urticae has been described by Gasser (1951), and its life history by Laing (1969b). *T. occidentalis* has been described by Chant (1959), and its life history by Laing (1969a).

In Figure 2 the 95% confidence interval of the number of prey killed per hour observed by Kuchlein and co-workers has been plotted against prey density. In spite of the high variance some conclusions can be drawn from this figure. The functional response curves show a steep rise and an abruptly decreasing slope at the lower prey densities

5

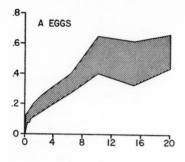

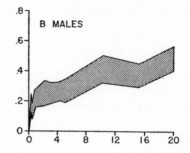

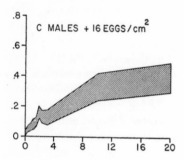

Fig. 2 | Areas of the 95% confidence interval of the number of prey killed per hour by a young female *Typhlodromus occidentalis* at different numbers per cm^2 of *Tetranychus urticae* (Kuchlein, unpubl. data).

below two prey per cm^2. They do not level off to a plateau immediately, but there is a rather linear rise between two and ten prey per cm^2. For prey males the curve may have some irregularities in the lower prey density range, and its slope may increase again at prey densities higher than fifteen males per cm^2.

2.2 Components of prey behaviour

In the system to be modelled two developmental stages of the prey species are introduced: eggs and adult males. The eggs are immobile and are distributed randomly over the disk surface. The males show a more complicated behaviour. They walk on the leaf surface with a certain velocity, or they feed on the epidermal cell contents in a typical, upwards-stretched posture. When walking the males tend to gather at certain places, and they produce an invisibly fine cover of webbing over the leaf surface with silk threads, 0.03-0.06 μm thick, from the mouth parts. The proportion of time the males spend walking will be termed the male activity.

2.3 Components of predator behaviour

When searching the predator vivaciously walks over the leaf surface with a certain velocity. It has some preference for the prominent parts, like veins and hairs. Periods of walking alternate with periods of cleaning, resting, or ovipositing, usually passed at sheltered places near veins or under hairs. The proportion of time spent walking during searching periods will be termed the predator activity.

The predator feels around with its 0.1-mm long front legs. Potential prey is perceived by contact with these front legs, which implies a short and constant maximum distance of perception. Each contact between prey and front legs marks an encounter between predator and prey. The number of encounters per time unit will be termed the encountering rate. An encounter results in a capture, if the prey is seized and killed. Not all encounters result in a capture. The proportion of captures will be termed the success ratio. A resting predator will encounter active prey males only, while a walking predator will contact resting males and prey eggs as well. As explained in the next section the encountering rate and the success ratio will vary with the prey species and with the encountering situation, and hence, have to be considered separately for eggs and males and for resting or active predator and males. If a prey is seized, its shell or integument is punctured with the mouth parts and the contents are sucked, until the prey has shrivelled or the predator is satiated. As soon as the prey has been punctured, it is doomed to die due to desiccation, although prey males may escape and survive for a while. Thus a capture is followed by a period of handling the prey, which includes a period of feeding. The latter is interrupted frequently by periods of piercing and palpating. While handling a prey, the predator is only rarely distracted by other prey bumping into it, so captures during handling periods will be neglected.

2.4 A preconceptual model of the predator-prey system

After this general outline of the behaviour of predator and prey, it has to be concluded which components of behaviour and which state variables of the system determine the number of captures per time unit.

Further it has to be conceived, whether these relevant elements are related to each other. Subsequently, the relationships conceived have to be built into a simulation model.

The first two steps of this approach, the conception of the relevant elements and their relationships, are tentative and lead to a provisional idea of the system, a preconceptual model. The ultimate model has to describe a part of the real world to such an extent, that it explains and predicts only the significant aspects of the actual predation process. Therefore, the preconceptual model used as a scheme for further analysis has been restricted to the most obvious elements and relationships. During the study the model has been extended by degrees until the output of the simulation model conformed to independent empirical results. The final presumptions made will be inferred in the next sections, which discuss the relevant elements and their mutual dependency. These presumptions originated partly at the instigation of the results of simulation, but for the sake of simplicity we will consider them as a starting point for the construction of the simulation models derived in the next chapters.

The feedback to a preconceptual level urges the investigator to gain insight into actual processes. This is an important advantage of the systems approach.

2.4.1 *The encountering rate*

The number of captures per time unit (predation rate) depends on the number of encounters per time unit (encountering rate) and the success ratio. The encountering rate is proportional to the velocity of the predator relative to the prey, so it will be influenced by the engagement of the predator in handling and searching and by the predator and prey activity. The relative velocity can be approximated by the square root of the sum of squares of predator and prey velocity (Skellam, 1958). The encountering rate is determined also by the preference of predator and prey for the same parts of the area. To quantify this it will be helpful to define the coincidence in space as the number of encounters per prey per unit of length covered by the predator relative to the prey. Then the encountering rate equals the product of the prey density, the relative velocity, and the coincidence in space.

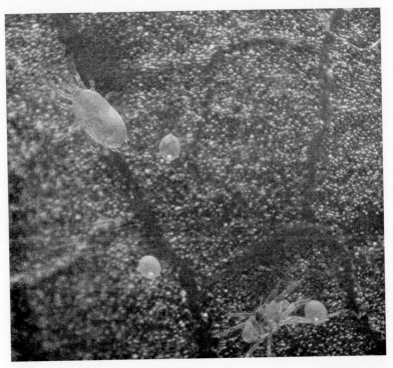

Plate Ia │ A *T. occidentalis* female on the lower surface of a bean leaf with a male and eggs of *T. urticae*. (50 ×)

Plate Ib | The attack on a prey male.

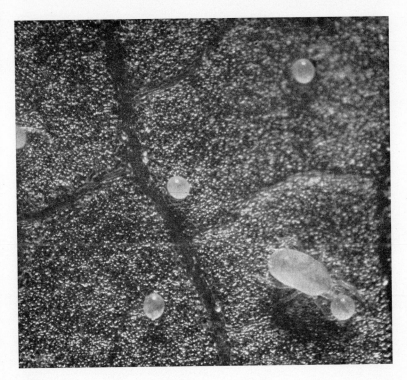

Plate Ic | The attack on a prey egg.

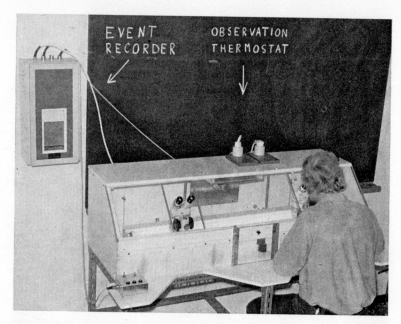

Plate II | Equipment for continuous observation.

2.4.2 *The success ratio*

The intensity of attack stimuli emanated from the prey varies with the prey species and the prey activity. For example, walking prey males evoke pursuit by *T. occidentalis*, which tends to increase the responsiveness of this predator. This effect is counteracted to some extent by a greater chance of escape. The responsiveness to prey of *T. occidentalis* also differs with its state of activity. Furthermore it can be assumed, that the success ratio is affected by the predator's hunger, and that it approaches zero when the predator is handling prey.

The frequency of encounters may induce response waning, which phenomenon has been termed inhibition by prey by Holling (1966). Inhibition of the predator by prey will appear as a direct influence of prey density on the success ratio.

Finally some influence of the density of the webbing cover cannot be excluded.

2.4.3 *The length of the handling periods*

The periods of handling prey include feeding periods. The time spent feeding on a prey will depend on the prey species, on the hunger of the predator, on the density of active prey disturbing the predator, and perhaps also on the density of the webbing cover, the shelter of which often seems to be appreciated by the predator.

The time spent in other components of handling, like palpating and piercing, will depend on the prey species and the hunger of the predator also. It may almost be proportional to the feeding time, because feeding is often briefly interrupted and subsequently continued on other parts of the prey, especially when the prey is male. While palpating and handling a prey in this way the predator is sensitive to disturbance by active prey and the presence of webbing, which interferes with its turning the prey. Therefore, the time spent handling, but not feeding, may also be related to the density of active prey and the webbing cover.

2.4.4 *The coincidence in space of the predator and the prey*

The number of encounters per prey per unit of length covered by *T. occidentalis* relative to the prey is determined by the size of the prey and the conformity of the spatial preference. The size is fixed for each sort of prey, but the spatial distribution and preference of the predator and the prey males is variable, and may be related to many components of behaviour and to state variables.

When walking, *T. occidentalis* to some extent prefers the prominent parts of the leaf surface, while walking males of *T. urticae* confine themselves to webbing covered areas and tend to form aggregations, which repel the predator. Resting predators prefer sheltered places near veins, but feeding males are dispersed more widely than walking males. Webbing cover restricts the freedom of movement of the predator, and it may isolate the prey eggs and the feeding males. These phenomena suggest that the conformity of the spatial preference of both the predator and the prey depends on the locomotion activities, and on the densities of the males and the webbing cover. There may be additional influences of the predator's hunger and the velocities of locomotion. When running fast the animals follow a straight path, which causes the predator to run along the veins, and scan a smaller area per unit of length covered than at lower speed.

2.4.5 *The locomotion activity of the predator and the prey*

The predator activity refers to the searching periods only. It is determined by the mean period of walking and the mean period of resting. The mean period of walking may be influenced by the density of the webbing cover, by the hunger of the predator, by its velocity, and by the density of the active prey males, which sometimes frighten the predator. The mean period of resting may be influenced by the hunger of the predator, by the density of the webbing cover, and by disturbance through active prey males. The disturbance will be defined as the proportion of encounters with active prey males, which result in the reactivation of the predator. This disturbance may depend on the predators hunger, on the density of the webbing cover, and on the density of active males, because frequent encounters may induce habituation to contacts with males.

The tendency of the prey males to aggregate will be accompanied by an influence of their density on their activity. The preference of the males for webbing covered areas may be a reason for a high walking activity at low densities of the webbing cover, while at high densities of the latter the males are more or less separated from their feeding sites, which also enhances locomotion. It seems reasonable to assume, that the male activity depends on the densities of the males and the webbing cover.

2.4.6 *The locomotion velocity of the predator and the prey*

The velocity of the walking predator may be related to the animal's hunger and activity, and to the density of the webbing cover. It is reduced by inspection of prey encountered, and increased by frightening encounters with active prey. This may be expressed as a relationship with prey density and prey activity.

The legs of the prey males are adapted for walking on webbing cover so the density of this cover may influence the velocity of the males. Because mutual contacts or olfactory stimuli effect aggregation, this velocity may be negatively correlated with the male density. Also, the activity of the males may affect the mean velocity of the walking males. The males reduce their velocity when they prepare for feeding. At low activity values most of the slowly walking animals have settled down, and the mean velocity of the walking males is high.

2.4.7 *The state variables of the system*

The state of the system at any one moment is determined by (1) the predator's engagement: whether it is walking, resting, or handling prey, (2) the actual densities of the prey species, (3) the hunger of the predator, and (4) the density of the webbing cover. These state variables (see p. 3) determine all other variables considered in the previous sections. These other variables are auxiliary, because their purpose is to elucidate the relationships between the state variables and the output variables, such as the number of prey killed.

The state variables are determined by their initial value and their rate of change. They have to be given a concrete, measurable form, and their rate of change has to be considered. The variables mentioned will be discussed one by one.

The engagement of the predator. The predator may be engaged in walking, resting, or handling prey. As mentioned in Section 1.2 the state of the system will be considered to be constant during the time increments used for simulation. As the mean periods of walking and resting are short (both about three minutes) and the periods alternate in an irregular, unknown pattern, it will be difficult to determine the searching predator's state of engagement for each time-step (about one minute). For the sake of simplicity it will be assumed that, according to its activity, a predator walks as well as rests during each searching period, and that this condition holds true for the time-intervals used for simulation assuming the periods of walking and resting to be very short. Then the state of engagement of a searching predator is accounted for by the activity of the predator, which is an auxiliary variable. To define the state of engagement during different intervals, it will be sufficient to distinguish between searching and handling prey only. This can be described by a discrete variable, being zero when the predator is searching, one when it is handling a prey egg, and two when it is handling a prey male. Most of the time the rate of change of this state variable will be zero. It will be one over a time-interval, if during this interval the predator catches a prey egg. It is minus one per interval, if the egg is abandoned. The rate of change is two and minus two per interval, respectively, when the prey is a male. The rate of change is determined by the probability of a capture or abandonment to occur during a certain interval. The capture probability can be computed from the encountering rates and the success ratios. The probability of an abandonment will depend on the predator's hunger, and on the other determinants of the handling period.

The actual prey density.   Each time a prey is captured the actual number of prey on the leaf disk is reduced by one. After fixed time-intervals of thirty minutes new prey is substituted to replenish the prey number. The mean actual number, however, will be lower than the initial number. The rate of change of the actual prey number during a small time-interval depends on the probability of a capture, and on the frequency and timing of prey replenishment.

The hunger of the predator. Hunger is a physiological condition, which is measurable by its effect on behaviour only. To measure it, it may be better to consider the opposite of hunger: the degree of satiation. Satiation can be measured as the difference between the amounts of biomass ingested and recognizably removed from the gut by absorbtion and defecation (together to be termed digestion for the sake of brevity). This is an usable measure, when ingestion can be determined in absolute units of food and digestion is considered as a relative disappearance of food. The amount digested should be indicated by other, measurable components of behaviour. This measure of satiation will be termed the gut content. It represents an internal condition, rather than a component of behaviour, e.g. the amount of food the animal is willing to accept in feeding experiments. Since satiation, and hence hunger, are closely related to the gut content defined in this way, this measure will be used as a convenient state variable.

The rate of change of the gut content is determined by the rate of ingestion, and by the rate of digestion. The rates of ingestion and digestion both may depend on the prey species and on the gut content itself. Because only liquid food is taken, it will be assumed that egg and male material are digested at equal rates. The determination of the digestion rate requires a priori knowledge of the relationship between the gut content and the predator's behaviour. A solution to this is included in the analysis of the experiments, which have been carried out to confirm and to describe the relationships assumed in this chapter (Section 3.4).

The density of the webbing cover. The thin silk threads produced by the prey males could not be made visible with the microscope used. Moreover, the density of the treads could not be measured without spoiling the system, which made it difficult to measure components of behaviour simultaneously. Therefore, it was impossible to relate the rate of increase of the webbing density to other system elements directly, and it proved to be necessary to introduce another state variable with a known rate of change, from which the density of the webbing cover can readily be computed.

Obviously, the density of the webbing cover is related to the total distance walked by the prey males on the leaf disks. This distance is derivable from the number of males present, the locomotion activity and velocity of the males, and the time elapsed since the preparation of the leaf disks. Its rate of change is the product of the first three of these factors. Provided, that the relationship between the density of the webbing cover and the total distance covered by the males can be obtained experimentally, the first may be considered as an auxiliary variable, and the latter may be used as a suitable state variable.

Table 1 Structural relationships, indicated by ×, in a preconceptual model of the predator-prey system. Underlined: state variables. Encircled: obvious relationships (see text). The names of the variables are listed in Appendix IV.

Dependent system elements	Determinants

```
                Determinants
                P        A A   R D V V    D D R G      D
                R E S    C C A E I E E P  E E E U I D E
                E N U H F C T T C S S L L S  N N P T N I N D
                D C C T T O M P T T M P P    E M D C G G W I
                R R R I I I A R I I U A R E  G A E O R R E S
                T T T M M N L E M M R L E C H G L L N T T B T

  PREDRT        ⊗ ⊗
   ENCRT             ⊗ ⊗ ⊗       ⊗ ⊗     ⊗
   SUCRT             × ×           × ⊗ × ×       ×             ×
    HTIM         ×    ×             ×         ×     ×       ×
    FTIM              ×             ×         ×     ×       ×
    COIN            × ×         × × ×       ×     ×       ×
  ACTMAL                              ×           ×
  ACTPRE              ⊗ ⊗
   ACTIM            ×             ×       ×     ×       ×
  RESTIM        ×             ×           ×       ×
  DISTUR            ×               ×     ×       ×
  VELMAL            ×                 ×           ×
  VELPRE           × ×             × ×     ×       ×
   PSPEC
       H        ⊗        ×           × ⊗     ×     ×       ×
    ACNE        ⊗                         ⊗
    ACNM        ⊗                         ⊗
  REPDEL
  GUTCON                                    ⊗ ⊗
   INGRT                          × ⊗         ×
   DIGRT                                    ×
  DENWEB                                            ×
    DIST             ⊗           ⊗           ⊗
```

3   The analysis of the structural relationships

A survey of the relationships expected in the mite system is given in Table 1. Some of these relationships are obvious, and can be expressed by simple mathematical equations. For instance, the predation rate equals the product of the encountering rate and the success ratio, and the predator activity equals the mean walking period divided by the sum of mean walking period and mean resting period. Such relationships are encircled in Table 1. The expression of the other relationships, however, cannot be inferred analytically because of lack of knowledge of the underlying mechanisms. The present chapter deals with experiments and their statistical inference, that have been carried out to confirm and to express these underivable relationships.

The aim of the experiments was the determination of dependent variables in situations with different known values of the independent variables. Even if the values of the latter cannot be selected arbitrarily, they have to be varied and measured simultaneously to obtain concomitant values of dependants and determinants. The rationale of the analysis applied is, that during prolonged continuous observation of the predation process with different prey densities the state variables and the components of behaviour will change sufficiently to permit the measurement or computation of different combinations of concomitant values.

To compute the changes of the gut content and the density of the webbing cover during the observation periods, knowledge is required about the rates of ingestion and digestion, and about the relationship between the density of the webbing cover and the distance walked by the males. Separate experiments have been carried out to study the ingestion rate and the webbing density. After the description of some general conditions, these experiments are discussed, and attention is paid to the determination of the digestion rate. Then the measurement of the locomotion velocity, the schedule of the continuous observation, and the elaboration of its results are described. The relevant relationships are formulated mathematically and summarized in a conceptual model.

3.1 Materials used for observing and recording the predation process

The system described in Section 2.1 was observed under a binocular microscope with magnifications 6, 12 and 25 times. Two microscopes were placed in a cabinet with the oculars protruding through the glass front (Plate II). The microscopes could be controlled manually through elastic slots in the front. In the cabinet a constant temperature of 27 °C was maintained, and a fluorescent tube illuminated the inside. The average humidity in the thermostat was 52% r.h. Measurements with a copper-constantan thermocouple and a constantan-manganese thermocouple psychrometer (design Stigter/Jansen) indicated a temperature of 25.1 °C and a relative humidity of 85% in the 0.5-mm thick air layer immediately above the surface of the leaf disks. This deviation of the microclimate in the layer, where the mites live, is probably due to evaporation from the wet cotton-wool and the leaf's stomata.

A 10-channel event recorder, with two switchboards each for five channels, completed this outfit for two observers. For each observer there were two on/off-switches and three push-buttons for control of the pens, which write a line in one of two optional positions on the moving paper tape of the recorder. With this combination different continuous situations as well as discrete events could be recorded chronologically.

3.2 Standardization of the predator

Because predator behaviour may vary with age, as for example oviposition (Kuchlein, 1966), most experiments were done with predators of the same age. To minimize the variance of behavioural components, all predators were treated in the same way. They developed on bean leaves with an abundance of *T. urticae* at all developmental stages at 27 °C and 70% r.h. Female deutonymphs were collected from this stock and transferred to other leaves with food, where they moulted overnight. Next day the adult females were transferred to leaves without food with an equal number of adult males of *T. occidentalis*. After two days of starvation, during which the mites became flat and translucent with an assumingly zero gut content, the predators were ready for use in experiments. These mated, two-day-old females were used throughout, except for experiments with ovipositing predators. The standard predators do not oviposit for at least 24 hours, even when fully fed.

3.3 The predator's ingestion rate

As even the smallest prey, the egg, measures in volume about 1/5th that of the predator, captured prey individuals will normally not be ingested completely; only starved predators empty eggs completely. Therefore the number of prey captured cannot serve as a measure of ingestion. It is better to consider feeding time as a measure of ingestion. It was suggested by Haynes & Sisojevic (1966), that sucking predators like spiders may have a constant ingestion rate when they are feeding. If this holds for *T. occidentalis*, the amount of food ingested would be proportional to the feeding time recorded during continuous observation.

To examine the relationship between ingestion and feeding time, standard predators were fed with radioactive eggs and males. After different feeding periods, ranging from one to eight minutes, the biomass ingested was estimated from the amount of radioactivity in the predator. To measure the extremely small quantities involved, the following method was applied.

Three seedlings of dwarf bean were cut when they had formed two pairs of leaves. Under room conditions the three stems were placed in 10 ml distilled water, containing the radioactive isotope ^{32}P to a total amount of ten microcurie. After absorption, the solution was replenished with tap water. Two days afterwards, unmated pre-adult *T. urticae* females (of the deutochrysalis stage) were attached to the underside of the lower leaves in cages of plastic and nylon gauze. The cages were about two-cm wide and glued to the leaf. Two cages were used per leaf, each containing three females. After moulting, the un-mated females started feeding and produced eggs, from which only males emerged. On the tenth day after the introduction of the mites, radioactive eggs and males were collected. On the tip of a brush this prey was offered to standard predators on empty leaf disks in the observation thermostat. The predators were allowed to feed for a certain time, which was measured with a stopwatch. Feeding is indicated by the visible movement of particles in the gut.

Double feedings were used to compare the ingestion during feeding on one and two prey individuals for equal periods, revealing the difference between the possible influences on the ingestion rate of satiation and the decreasing availability of the prey's food content. A second prey, if it was offered immediately after the removal of the first when the predator was still feeling around for the prey, was readily accepted. An equal number of predators were allowed to feed for a certain period on one or two prey individuals. In this way batches of predators were obtained for each feeding period used. The number per batch differed for the prey species used, being five for eggs and three for males. Immediately after feeding the predators were killed in the vapour of ethyl acetate and transferred per batch to small, covered glass vials. To relate the amount of radioactivity to the number of radioactive prey, other vials were filled with one to five radioactive eggs, or one to three males. After addition of 0.2 ml perchloric acid and 0.4 ml 30% hydrogen peroxide, the vials were incubated for 30 min at 80 °C. The resulting clear solution was cooled and 7 ml 2-methoxyethanol (cellosolve) and 10 ml of a 0.6% toluene solution of 2,5-diphenyloxazol (PPO) were added. The vials were shaken, placed in a liquid scintillation counter and chilled for half an hour. After adjustment of the apparatus for the highest sensivity, the number of photons radiated from the solution per minute was counted during ten minutes. The photons are generated by collisions between the PPO molecules and neutrons, which are emanated by the ^{32}P. Therefore, the number of counts per minute is a measure of the amount of ^{32}P in the solution.

The experiments were repeated seven times with eggs and eight times with males. The mean number of counts per minute of empty vials and the vials containing radioactive prey are plotted in Fig. 3 against the number of prey in the vials. The straight lines, obtained for both prey species by linear regression, were used to compute the amount of prey ingested by the predators from the number of counts per minute of the batches. On the assumption, that the ^{32}P is ingested proportionally, the curves of Fig. 3 can be used inversely to obtain the amount of prey ingested in prey number. Because we use a relative measure,

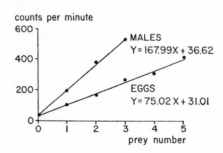

Fig. 3 | Gauge relating the number of counts per minute in the liquid-scintillation method to the number of radioactive prey (prey equivalents).

we will express this as the number of prey equivalents. The mean number of prey equivalents ingested per predator in the different feeding periods used is given for eggs and males in Fig. 4.

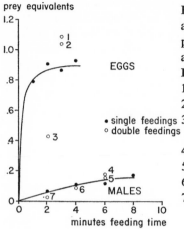

prey equivalents

minutes feeding time

Fig. 4 | The relationship between the amount of food ingested from labelled prey (expressed in prey equivalents) and feeding time in starved predators. Double feedings:
1: 2 + 1 minutes on labelled eggs
2: 1 + 2 minutes on labelled eggs
3: 2 minutes on an unlabelled male, 2 minutes on a labelled egg
4: 2 + 4 minutes on labelled males
5: 4 + 2 minutes on labelled males
6: 2 + 2 minutes on labelled males
7: 2 minutes on an unlabelled egg, 2 minutes on a labelled male.

The results of these feeding experiments show, that the ingestion rate of *T. occidentalis* is not constant. It decreases as feeding proceeds. This may be due to the exhaustion of the food content of the prey. When, however, the predators could feed liberally, as in the case of the double feedings, the decrease of the ingestion rate was still apparent. In the seven experiments with one predator feeding on two prey, it always ate less than did two predators, feeding on one prey each for an equal sum of time. This would have a probability of 0.008, if feeding itself did not effect the ingestion rate. It is therefore concluded, that the ingestion rate depends on the gut content.

To describe the relationship of the ingestion rate with the gut content mathematically, we assume that the ingestion rate is proportional to the difference between the actual gut content (A) and the maximum gut content (M). Then we can write:

$$dA/dt = c(M - A) \tag{3.1}$$

where c is a positive constant and t is the feeding time.

This integrates to:

$$A = M(1-e^{-ct}) \qquad (3.2)$$

in which e is the base of the natural logarithm.

To measure M and c in Eqn (3.2) for eggs and males, known values of A and t from two experiments, in which the predators could feed freely, were substituted in the formula. For eggs $t = 1$ min and $t = 3$ min were chosen, for feeding on one and two prey, respectively. For males these values were 2 and 6 minutes. The values of A can be obtained from Fig. 4. The A values corresponding with the lower t values were obtained with the regression curves through the points for single feedings, which are derived below. For the double feedings the average A values were used. After substitution we have for eggs:

$$0.81 = M(1-e^{-c})$$
$$1.06 = M(1-e^{-3c})$$

These equations were solved for M and c: M equals 1.08 egg equivalents and c equals 1.32 min^{-1}.

For males we have:

$$0.066 = M(1-e^{-2c})$$
$$0.169 = M(1-e^{-6c})$$

M equals 0.56 male equivalents and c for males equals 0.059 min^{-1}. Presumably the maximum gut content is the same for feeding on eggs and males, so 0.56 male eq. equals 1.08 egg eq. Now it is possible to convert male eq. to egg eq., the latter will be used consistently as the unit for biomass consumed. Thus we have:

$$A = 1.08(1-e^{-1.32t}) \text{ egg eq.} \qquad (3.3)$$
$$A = 1.08(1-e^{-0.059t}) \text{ egg eq.} \qquad (3.4)$$

for feeding on eggs and males respectively, when the initial gut content is zero. If the latter is in general A_0, the formulae can be generalized to:

$$A = 1.08 - (1.08 - A_0)e^{-1.32t} \text{ egg eq.} \qquad (3.5)$$
$$A = 1.08 - (1.08 - A_0)e^{-0.059t} \text{ egg eq.} \qquad (3.6)$$

When the prey content is not exhausted, these formulae will give a good description of the ingestion during definite feeding periods, because the majority of the periods observed are shorter than the highest

20

t-values used for the determination of the formulae.

The equations (3.5) and (3.6) will hold for predators, which have more than, say, 0.2 prey equivalent in their gut. However, when a starved predator feeds on a first prey, feeding is continued for a long time, starting with a high ingestion rate. Then, ingestion is more inhibited by the limited food content (*L*) of the prey than by satiation, although the predator continues sucking. This process is less gradual than the influence of satiation alone, since it reveals itself mainly when A reaches L. It can be modelled approximately by assuming, that A is proportional to t and $L-A$:

$$A = at(L-A) \tag{3.7}$$

where a is a positive constant. By making A explicit we have:

$$A = Lt/(a'+t) \tag{3.8}$$

where a' is another constant. To determine L, the remains of completely emptied radioactive prey were collected and their radioactivity was measured. One minus the biomass of the remains expressed in prey equivalents gives L. In this way L was found to be 0.94 egg eq. for eggs and 0.67 egg eq. for males. By fitting the formula (3.8) to the points of single feedings in Figure 3, all referring to starved predators feeding on a first prey, the relationship between ingestion and feeding time for the first prey was found to be:

$$A = 0.94t/(0.16+t) \text{ egg eq.} \tag{3.9}$$

for eggs and

$$A = 0.67t/(9.09+t) \text{ egg eq.} \tag{3.10}$$

for males.

These hyperbola are sketched in Fig. 4. Although the model (3.7) was chosen tentatively, the hyperbola seem to describe the relationships quite well.

In the analysis of observation records for standard predators the gut content after feeding on the first prey will be computed with the formulae (3.9) and (3.10). All subsequent values of the gut content after feeding will be computed by (3.5) and (3.6) because the gut content will remain too high for the rest of the observation period to enable the predator to utilize the next prey completely.

21

3.4 The predator's digestion rate

Absorbtion and defecation will decrease the gut content continuously. An exponential decrease is presumed, because it is plausible to suppose, for an animal taking liquid food and having a gut with a large surface in relation to its volume, that the rate of decrease of the gut content is proportional to the amount of food present. Then the rate of digestion, as we may call it, is represented by:

$$dA/dt = -bA \tag{3.11}$$

in which A denotes the actual gut content, b a positive constant and t time. If A_0 is an initial gut content, Eqn (3.11) integrates to:

$$A = A_0 e^{-bt} \tag{3.12}$$

This assumed proportionality of the digestion rate and the gut content is not completely imaginary. Such a relationship has been found by Holling (1966) in the praying mantid *Hierodula crassa*. Nakamura (1968) found an almost linear relationship between the ingestion rate and the digestion rate of the spider *Lycosa pseudoannulata*, while Mukerji & LeRoux (1969) found the same for the ingestion rate and the daily weight increment in nymphs of the bug *Podisus maculiventris*. It seems rather common in invertebrates, that for a wide range of the ingestion rate the food disappears from the gut at the same rate as it enters it. This implies that a food particle ingested is always evacuated from the gut at more or less the same rate; otherwise the gut would be filled to its maximum capacity at a low ingestion rate. Theoretically, the relationship (3.12) can be confirmed for *T. occidentalis* by a comparison of the ingestion of radioactive prey by predators with a known initial gut content, which have been deprived of food for different time-intervals. It would be possible to use the formulae (3.5) and (3.6) to calculate the gut content after the digestion intervals (A_0 in these formulae), and to check the linearity of the relationship between its logarithm and the digestion interval. Because *T. occidentalis* hardly responds to prey offered to the animal when its gut is not completely empty, this method has not been attempted. Instead it is assumed that by choosing the right value of b in the formula (3.12) this formula will give a good description of the digestion process.

As argued in Section 2.4.7 digestion will be measured by its effect on the predator's behaviour, which presupposes knowledge about the influence of the gut content on behaviour. Since it is intended to determine the relationship between the behaviour and the gut content (see p. 12), this reasoning seems to go round in circles. To solve this an iterative procedure was applied. This procedure is commonly used to solve two equations with two unknown parameters.

The predators used for continuous observation (see Section 3.7) were removed from the leaf disks at the end of the observation period and were deprived of food for one night. Next morning they were released on the same disks again, and their behaviour was recorded until they captured a prey and abandoned it after feeding. The components of behaviour considered are the success ratio and the handling time. The parameter b in Eqn (3.12) was approximated by repeating the next sequence of computations until b was constant:

1 Estimate the digestion parameter b and compute the changes of the gut content in the course of the predation processes observed with the formulae derived for ingestion and digestion $(A. = 0)$.
2 Relate the behavioural components to the gut content.
3 Compute the gut content after one night of digestion.
4 Compare the behaviour observed after one night of digestion with the behaviour to be expected from the relationships obtained and the gut content computed.
5 Improve the estimate of the digestion parameter b. Repeat the computations 1-5, if there is a difference between the behaviour observed and the behaviour expected.

To start this procedure, the parameter b was estimated roughly.

It was considered that the predator can produce up to four eggs per 24 hour when food is abundant (Kuchlein, 1966). Although standard predators do not lay eggs, their digestion rate may be assumed to be of the same order as that of ovipositing females, because they have to build up the oocytes for egg production. Even if they require less energy, the amount of food needed will be controlled by the success ratio rather than by the digestion rate (see Section 3.8). Predator eggs have about 4/3 the volume of prey eggs, so the production of four predator eggs requires the digestion of at least 16/3 prey eggs. The production of eggs will utilize most of the food consumed, but some energy is used for other activities. To make a rough estimate, it can be assumed that the predator will digest eight prey egg equivalents per 24 hour, when its gut is constantly filled to its maximum of 1.08 prey egg equivalents.

Then Eqn (3.11) reduces to:

$$dA/dt = -1.08b$$

For $dA/dt = -8/24$ egg equiv., b follows from this equation and was estimated as $0.309\ h^{-1}$.

The range of the gut content from 0 to 1.08 egg eq. was divided into twelve classes: one zero class, ten classes comprising 0.1 unit each, and one class ranging from 1.00-1.08 egg eq. The values of the components of predator behaviour were computed for each of the gut-content classes as described in Section 3.7. By plotting the mean values of the success ratio and the handling time over the medial values of the gut-content classes, relationships between these components of predator behaviour and the gut content were obtained.

By correcting b and repeating the computations the best estimate of this parameter was found to be $0.435\ h^{-1}$. The gut content is halved by digestion in about 1.6 hours.

3.5 The relative density of the webbing cover produced by the prey

The thin silk threads produced by the prey males are invisible with the magnifications used. For the indirect determination of the density of the webbing cover some experiments were carried out to relate this density to the readily computable distance covered by the males.

Leaf disks with webbing were sprayed with a fine grained yellow powder, which is normally used as a fluorescent. The grains tended to stick to the silk threads. As an indirect measure of the webbing density the proportion of grains on the leaf sticking to threads was used. It has a value between zero and one.

It can be expected, that this relative density of webbing is related to the total distance walked by the prey males. This distance is a simple function of the number of males present, their locomotory activity and velocity, and the time elapsed since the preparation of the leaf disk. If the males walk at random over the leaf disks and produce threads wherever they are, then the increment of the relative webbing density (W) with the distance covered (D) decreases only by the gradual covering of the surface; uncovered places becoming increasingly scarce. This may be represented by:

$$dW/dD = c(1 - W) \tag{3.13}$$

where c is a positive constant. This integrates to:

$$W = 1 - e^{-cD} \tag{3.14}$$

with e the base of the natural logarithm. After some transformations we have:

$$\log(-\ln(1-W)) = \log(D) + \log(c) \tag{3.15}$$

This equation can be represented graphically by a straight line with a 45° angle with the abscissa.

Spider mite males were allowed to walk for different periods of time on leaf disks under standard conditions in the observation thermostat. The locomotion velocity of forty males was measured repeatedly at the end of each time period by the methods described in Section 3.6, while the locomotory activity was determined as the average proportion of walking males. Table 2 gives the number of the experiments, the mean values of the distance covered (D) and the relative density of webbing (W) as measured by the grain method. In Fig. 5 $\log(-\ln(1-W))$ is plotted against $\log(D)$. As shown, the existence of a linear relationship is refuted. Since a curve through the points tends to level off, it may be concluded that the webbing threads are not deposited at random. The production of webbing is gradually reduced, or the males tend to walk along common pathways. Quadratic regression, which gives a significantly better fit than linear regression ($P < 0.01$), yields:

$$\log(-\ln(1-W)) = -0.133(\log(D))^2 + 0.879 \log(D) - 1.607$$

This can be written as:

$$W = 1 - e^{-0.0247 D^{(-0.133 \log(D) + 0.879)}} \tag{3.16}$$

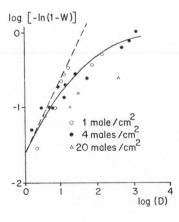

Fig. 5 | The relationship between $\log(-\ln(1-W))$ and $\log(D)$, where W is the relative density of the webbing cover and D is the total distance walked by *T. urticae* males.

25

Table 2 The distance covered in metres (*D*) and the density of the webbing cover (*W*) in experiments with *T. urticae* males on a 5 cm² area. DENMAL is the density of the males in number/cm², n is the number of replicates, *t* is the time elapsed in hours, and s_W is the standard deviation of *W*.

DENMAL	n	t	D	W	s_W
1	3	0.5	2.19	0.03	0.02
1	3	1	4.27	0.08	0.03
1	3	2	7.38	0.11	0.01
1	3	3	9.88	0.20	0.03
1	3	4	11.89	0.24	0.03
1	3	6	16.69	0.29	0.02
1	3	26	76.05	0.32	0.10
1	3	43	133.43	0.40	0.06
4	3	0.08	1.64	0.05	0.01
4	3	0.17	2.69	0.09	0.01
4	3	0.33	5.74	0.09	0.02
4	3	0.38	4.98	0.10	0.02
4	3	0.67	10.51	0.17	0.02
4	3	0.75	12.15	0.13	0.03
4	3	1	12.60	0.18	0.01
4	3	2	22.96	0.24	0.05
4	3	3.75	53.76	0.33	0.05
4	3	24.5	383.14	0.46	0.03
4	3	50	792.78	0.53	0.07
4	3	74	1197.30	0.63	0.01
20	2	0.5	18.18	0.10	0.02
20	2	1	30.30	0.16	0.04
20	2	6	218.16	0.22	0.02

Equation (3.16) will be used in simulation models as the relationship between the webbing density and the feasibly estimated distance covered. For high male densities the equation seems to overestimate the webbing production (see Fig. 5). This may be due to aggregation at high male density, which causes an irregular distribution of the webbing cover. For the time being this phenomenon will be neglected, because there is too little information to give a good quantitative description.

3.6   The measurement of the locomotion velocity

The speed of locomotion of predators and prey males can be measured directly during observation. Two methods have been used. In the beginning, the walking path of the animal observed for one minute was drawn to scale in a sketch of the leaf disk. From this path the locomotion velocity can be derived with a curvimeter. This method proved to be very laborious and liable to scaling errors.

Later another method was applied to obtain quick estimates of the velocity during continuous observation. The grid of squares of an ocular micrometer was superimposed on the twelve times magnified leaf disk, the squares of the grid being small compared to the animals observed. A number (n) of grid squares traversed consecutively by the walking animal during time t was counted (Fig. 6). The side of the squares could be measured with an object-micrometer: $s = 0.85$ mm. The velocity (V) was then estimated approximately by:

$$V = \mu_L ns/t$$

where μ_L is the average random linear traverse through a unit square ($s = 1$). Monte Carlo estimation (Appendix I) revealed a value of 0.709 as the mean of 5 000 random traverses of a unit square. The standard deviation of this sample was 0.364. The standard deviation of the estimate of the velocity V was $0.364\sqrt{n}\, s/t$, which is less than five percent of V if n equals 106. During the measurements the animals

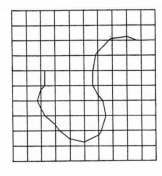

Fig. 6 | Path of a walking mite on a grid of squares, as indicated by a sequence of linear traverses.

were allowed to traverse at least 200 squares ($n > 200$). Each time ten squares were traversed, the observer indicated it on the event recorder, so that the number of squares traversed and the time elapsed could be read from the records of continuous observation.

During nine observations the velocity of standard predators was measured before and after the first feeding. The mean velocities were 1.15 and 1.12 m/h, respectively. No significant difference was found, and this strongly suggests, that the gut content has no influence on the predator's velocity.

3.7 Continuous observation of the predation process and the analysis of the records of events observed

The continuous observations of the predation process have been made on leaf disks with different prey densities (Table 3). One standard predator was released per disk. This predator could be observed continuously for several hours. The events recorded chronologically are:

1 initiation and termination of walking periods (on/off switch)
2 initiation and termination of feeding periods (on/off switch)
3 encounters with eggs, walking males, and resting males (push buttons), respectively.

During pauses in the observation the predators were transferred to empty disks. Captured prey individuals were replenished immediately. If present, the resting males were counted every quarter of an hour, and the locomotion velocity of the predator and a male was measured six times in the course of the observation period.

For each leaf disk the values of some variables were known, or could be calculated from direct measurements: the prey densities, the locomotion velocities, the mean activity of the prey males, and the density of the webbing cover. Their average values are given in Table 3. The system elements to be computed from the records of events as concomitant variables are the gut content and the remaining components of the predator's behaviour.

The records are a chronology of events in the predation process observed, which were coded by the set of numerical values given in Table 4. The events indicating a change in the predator's engagement were labelled with their time of occurrence from the start of the observation period. In this respect the periods of handling prey were regarded as a

Table 3 Summary of observations made with standard predators on continuously observed leaf disks, with the mean values of the variables measured, the number of replicates (n), and the total observation time in hours (t). The names of the variables are listed in Appendix IV

DEN EGG	DEN MAL	DEN WEB	VEL PRE	VEL MAL	ACT MAL	n	t
1	0	—	1.39	—	—	6	27.7
4	0	—	1.30	—	—	6	29.3
20	0	—	1.43	—	—	6	27.7
0	0.4	0.01	1.33	0.73	0.78	3	2.2
0	1	0.08	1.07	0.83	0.60	6	21.1
0	2	0.04	1.01	0.55	0.76	3	1.9
0	3	0.05	0.99	0.55	0.73	3	2.1
0	4	0.17	0.91	0.66	0.59	6	18.1
0	10	0.09	0.73	0.32	0.72	3	1.5
0	20	0.30	0.64	0.56	0.65	6	18.8
4	1	0.08	1.07	0.83	0.68	6	22.7
4	4	0.17	0.91	0.66	0.68	6	22.7
4	20	0.30	0.64	0.56	0.67	6	17.9
4	4	0.09	0.97	0.67	0.59	24	10.9
4	4	0.30	0.86	0.65	0.73	24	10.7
4	4	0.50	1.09	0.90	0.77	24	12.6
4	4	0.55	0.89	0.84	0.76	24	12.4
4	4	0.56	0.68	0.78	0.80	24	12.1

sequence of feeding and non-feeding periods, which is terminated by the end of the last period. The sequences of codes and time labels were punched on 80-column Hollerith cards. A FORTRAN program, called ANCOWA, was developed to process the extensive data with a computer. The program changes the gut content at the end of each time-period, with the formulae derived for ingestion and digestion. At the end of each time-period Eqn (3.12) is used to account for the digestion during this time-period. When the predator abandons a prey, subsequently Eqns (3.9) and (3.10) are used to account for ingestion when the initial gut content is zero. When the gut content at

Table 4 Numerical codes of events, as used in the computer analysis of the observation records

Code	Time labeled	Event
1	+	end of complete walking period
2	+	end of incomplete walking period
3	+	end of complete resting period
4	+	end of incomplete resting period
5	+	restart of observation after pause
6	+	end of feeding period
7	+	end of handling but not feeding period
8	−	the predator abandons the prey
9	−	encounter between a walking predator and a walking male
10	−	encounter between a walking predator and a resting male
11	−	encounter between a resting predator and a walking male
12	−	encounter between the predator and an egg
13	+	disturbance of the predator by a walking male
14	+	capture of a walking male by a walking predator
15	+	capture of a resting male by a walking predator
16	+	capture of a walking male by a resting predator
17	+	capture of an egg
18	−	the predator got lost, read additional data for another predator and reset the gut content to zero

the moment of capture is not zero, Eqns (3.5) and (3.6) are used for ingestion. In the formulae for ingestion, the value of time is the sum of the feeding periods between capture and abandonment. Table 5 mentions the quantities, which are averaged by ANCOWA for the different gut-content classes mentioned in Section 3.5 (see p. 24). For information on the magnitude of these quantities and on the extensiveness of the observations the overall mean values and the total number of periods and encounters are included in the table. Obvious mistakes in the numerical records, such as nonexisting codes or the end of a walking period, when the prey is not abandoned, are indicated and located by ANCOWA. Because all possible errors in coding are detec-

Table 5 The quantities computed from the records of observations of continuously observed predators, with their overall mean values and the total number of the periods and encounters observed. The indications A A–A E refer to the different types of encounters (see Appendix IV)

Quantity	Dimension	Mean	Total number
average walking period	hour	0.045	2224
average resting period	hour	0.043	2373
the activity of the predator	—	0.520	—
average handling period per egg	hour	0.081	279
average feeding period per egg	hour	0.067	279
average handling period per male	hour	0.201	260
average feeding period per male	hour	0.134	260
disturbance of the predator	—	0.429	—
encountering rate A A per prey	hour^{-1}	2.066	4412
encountering rate A R per prey	hour^{-1}	2.195	2253
encountering rate R A per prey	hour^{-1}	0.894	583
encountering rate A E per prey	hour^{-1}	0.701	2964
success ratio A A	—	0.038	—
success ratio A R	—	0.034	—
success ratio R A	—	0.026	—
success ratio A E	—	0.097	—

table in this way, the results of the analysis are not corrupted by inevitable inaccuracies.

Tables 4 and 5 need some elucidation. Complete walking and resting periods are initiated and terminated by the predator, not by the experimenter. A capture appears in the experimental records as an encounter followed by a period of feeding, which lasts for at least thirty seconds when the prey is a male. For the definition of the other terms used see Section 2.4.

As a result of the analysis, numerous values of the variables mentioned in Tabel 5 were obtained for several classes or values of gut content, prey density, density of the webbing cover, prey activity, and locomotion velocity of the predator and the males. The coincidence in space was derived from the encountering rate and the locomotion velocities. A presentation of the class averages of each single factor has been omitted. This would have been too voluminous, and the influences of the factors are too complex for simple evaluation. The values were entered into multiple correlation and regression analysis, as described in Section 3.9.

3.8 A comparison of standard predators and ovipositing predators

To enable the computation of the gut content, the standard predators were conditioned so as to have an empty gut and not to lay eggs. In Kuchlein's experiments, the results of which have to be explained by this analysis, some predators did lay eggs. Egg production may affect several components of behaviour. To study this, some ovipositing predators were observed continuously on leaf disks with different prey egg densities, the results of which are given in Tables 6 and 7.

The ovipositing females were taken from the stock and adapted for 24 hours to the prey density used. Consequently, the observations on these predators were not started with a zero gut content. As a starting point for ANCOWA, the average gut content from experiments with standard predators and corresponding prey density was introduced as an initial value. The average gut content of the ovipositing predators did not deflect consistently from this value (Table 7).

As shown in Table 6, the predators are able to lay eggs for some time after the depletion of food, indicating the existence of a second reserve of energy apart from the gut content. The oviposition rates given in

Table 6 Summary of observations made on ovipositing predators on continuously observed leaf disks. The values of the prey egg density (DENEGG), the means of the velocity of the predator (VELPRE), the numbers of eggs laid per 24 hour (o_1), the numbers of eggs laid per 24 hour on disks without prey after one night of food depletion (o_2), the numbers of replicates (n), and the total observation time in hours (t) are given

DENEGG	VELPRE	o_1	o_2	n	t
1	1.24	1.14	0.33	3	6.2
4	0.92	1.69	1.00	3	6.4
20	1.25	2.25	1.33	3	5.6

Table 7 A comparison of the results of continuous observations made on standard predators, ovipositing predators, and predators studied by Kuchlein (unpublished data), at different prey densities

Variable	Number of prey per cm^2		
	1	4	20
eggs encountered per 24 hour by			
standard predators preying on eggs	76	202	1191
ovipositing predators preying on eggs	74	99	698
Kuchlein's predators preying on eggs	47	216	878
eggs captured per 24 hour by			
standard predators preying on eggs	15.7	19.6	28.0
ovipositing predators preying on eggs	23.2	30.0	34.3
Kuchlein's predators preying on eggs	3.2	6.2	13.2
males captured per 24 hour by			
standard predators preying on males	38.9	50.2	57.5
Kuchlein's predators preying on males	8.2	9.5	17.8
average gut content of			
standard predators preying on eggs	0.48	0.54	0.57
ovipositing predators preying on eggs	0.42	0.52	0.67
average locomotion activity of			
standard predators preying on eggs	0.37	0.30	0.31
ovipositing predators preying on eggs	0.34	0.32	0.21
average locomotion velocity (m/hour) of			
standard predators preying on eggs	1.53	1.69	1.61
ovipositing predators preying on eggs	1.24	0.92	1.25

Table 6 will be considered in the discussion on the numerical response (Section 5.4.1).

Table 7 compares the number of prey eggs encountered and the number destroyed in 24 hours by ovipositing predators and standard predators with non-zero gut content, for three densities of prey eggs. As indicated, the number of prey captured per 24 hours is increased by oviposition, although the encountering rate seems to be slightly reduced. Egg production seems to increase the predation rate by raising the success ratio. Some components of behaviour are compared in Table 7 also, showing that egg production reduces the locomotion velocity. This proved to be significant ($t_{21} = 3.61$).

With regard to the differences found between standard and ovipositing predators, it was to be expected that in Kuchlein's experiments more prey was captured per time unit than in the experiments with standard predators under the same environmental conditions. Table 7 presents some numbers of prey encountered and captured as recorded by Kuchlein (pers. commun.). Surprisingly, the predation rate is about four times as low in Kuchlein's experiments, mainly due to a lower success ratio. To obtain models, which simulate the functional response in these experiments, it will be necessary to account for the difference in success ratio. Although this difference cannot be explained, it seems most reasonable to multiply the success ratios derived in the present analysis by a factor 0.25.

3.9 Polyfactor analysis of the multiple relationships between the system elements

So far the analysis has provided rows of concomitant values of several variables. Most of these variables may be assumed to be dependent on some of the others, as indicated in Table 1. The aim of the next step is to define quantitatively the multiple relationships between these variables measured from the data obtained. Lack of knowledge of the underlying mechanisms forces us here to leave the mere deductive approach based on a priori assumptions championed by Holling (1963, 1964), and to follow a more descriptive approach, as outlined by Watt (1961, 1966, 1968).

The determinants tend to obscure each others influence, so that it is impossible to study these influences separately. To describe the multiple relationships, it will be necessary to use a multiple regression model. Because curvilinearity is a very interesting feature of the relationships studied, the regression models used must have sufficient terms with high powers of the determinants. To obtain manageable descriptive models of sufficient complexity and flexibility in a feasible way, an iterative procedure, as described by Ferrari (1952), was applied. Following Ferrari, this procedure will be called polyfactor analysis. As an example, the procedure will be worked out for the activity of the

34

prey males (ACTMAL) as dependent on the male density (DENMAL) and the density of the webbing cover (DENWEB). The values of the variables involved can be obtained from Table 3.

In the original method the dependant ACTMAL is plotted against one of the determinants, say DENMAL, and a smooth curve is drawn graphically with the least sum of squares of deviations (Figure 7A). This curve roughly describes the influence of DENMAL on ACTMAL, but the relation is biassed by the effect of DENWEB on ACTMAL, which is incorporated in the values of ACTMAL plotted. The curve may be used, however, to obtain a better description of the influence of DENWEB on ACTMAL. To this end the values of ACTMAL on the curve are subtracted from the values of ACTMAL measured, and their mean value is added. In this way the values of ACTMAL measured are corrected for the influence of DENMAL. The mean of the values on the curve is termed the correction level. The values of ACTMAL thus corrected are plotted over DENWEB, and another curve is drawn (Figure 7B). If there would have been another determinant, the corrected values of ACTMAL would then be corrected for the influence of DENWEB and plotted over this additional determinant, and so on.

After the first step there are two curves representing functions of the determinants: $f(DENMAL)$ and $f(DENWEB)$. After subtraction of the correction levels $L(DENMAL)$ and $L(DENWEB)$ the reduced functions are:

$$F(DENMAL) = f(DENMAL) - L(DENMAL)$$
$$F(DENWEB) = f(DENWEB) - L(DENWEB)$$

and an additive model is given by:

$$ACTMAL = L(DENWEB) + F(DENMAL) + F(DENWEB) \quad (3.17)$$

The influences of the determinants are still rather entangled in the functions obtained. To improve the crude model, ACTMAL is corrected for the influence of DENWEB, and other determinants if present, and plotted over DENMAL. A new curve presents an improved $f(DENMAL)$ (Fig. 7C). ACTMAL is corrected for the influence of DENMAL with this new curve and plotted over DENWEB, providing an improved $f(DENWEB)$ (Fig. 7D). This procedure may be repeated (Figs. 7E and 7F) until the curves do not change any more. Then the model (3.17) gives the best estimate of ACTMAL for any values of

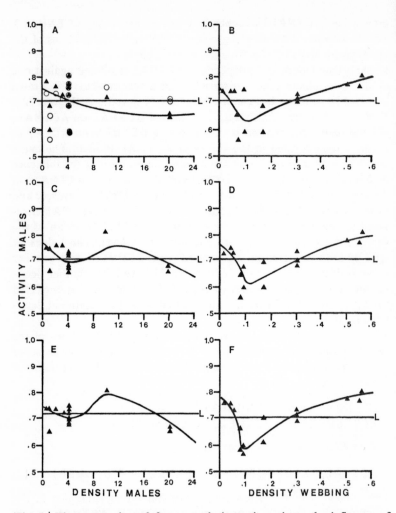

Fig. 7 | Three steps in polyfactor analysis, to investigate the influence of the number of *T. urticae* males per cm² and the density of webbing cover on the activity of the males. L is the correction level. The triangles are the values plotted, while the open circles represent corrected values.

DENMAL and DENWEB.

Figure 7 demonstrates that the sum of squares of the deviations decreases with an increasing number of iteration runs. In fact polyfactor analysis is a least square method to fit a regression model to a set of data points.

Drawing the curves by eye makes the method very flexible, but also very laborious. Moreover, the results are not easily incorporated into simulation models. In order to establish the numerous relationships involved in the present study, the manual plotting was replaced by either a compound linear regression method, or by fitting a third degree polynomial. In the first method linear regression on three points was applied repeatedly, taking the points in succession from low to high X-values and shifting one point at a time (Fig. 8). The middle points and the most extreme points were placed on the lines. By repeating this, smooth curves can be obtained. By making polyfactor analysis entirely computable in this way, it could be programmed in FORTRAN.

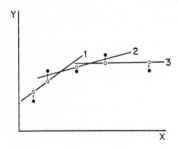

Fig. 8 | Linear regression as applied to obtain a smooth curve (represented by the open circles) through a set of points (solid circles).

A program was written, which continues iteration runs until the difference between all points of the functions in successive runs is less than one percent. The program takes into account the weight of the sample points, which is determined by the number of observations they represent. Another program was used to plot the resulting functions on a CALCOMP plotter.

Before the program for polyfactor analysis was applied for the description of the multiple relationships, an attempt was made to reduce the number of the relationships to be regarded. The rows of data were transformed to obtain a quasinormal frequency distribution of the variables, using the arcsine (x) for proportions and the $\ln(0.01 + x)$

for prey densities. Then the partial correlation coefficients of all pairs of variables, eliminating the influence of the other variables, were computed to study the linear dependency on the determinants. For the reader familiar with the use of matrix algebra, the derivation of the partial correlation coefficients can be described as follows:

With a computer program the correlation coefficients of all pairs of rows of transformed variables are computed and arranged in a correlation matrix. The elements of the inverse of the correlation matrix are divided by the square root of the corresponding elements on the diagonal. Then the partial correlation coefficients are obtained by reversing the sign of the resulting values of the elements, except for the elements on the diagonal.

Because curvilinearity of the relationships can interfere with correlation, this had to be taken into account. The rows of data were entered into the computer programs for polyfactor analysis, using the compound linear regression method. Seventy-one plots were obtained for the fifty non-encircled relationships of the components of behaviour in Table 1, including the success ratio and the coincidence in space for the different prey species and the activity of the predator and the prey. On the base of significant partial correlation and the shape of the functions plotted, thirty-two relationships were selected to be considered in further analysis and simulation. These relationships, with the statistics derived, are given in Table 8.

Polyfactor analysis as outlined above has an important disadvantage. In case of correlation of the determinants all influence on the dependent variable is attributed to the determinant with the highest effect. The method is much improved, when the determinants are replaced by linear combinations of the variables, which are uncorrelated. Therefore, the determinants listed in Table 8 were replaced by orthogonal linear combinations. Combinations having a sum of squares equal to one (orthonormal linear combinations) were used, because they could be obtained with a computer program most conveniently. They also are uniformly scaled to provide the smallest rounding off errors in polyfactor analysis.

The derivation of orthonormal combinations is illustrated by Fig. 9 for the two variables DENMAL and DENWEB used in the example. Table 3 gives the fifteen concomitant values both of DENMAL and DENWEB. These rows of values represent two vectors in fifteen-dimensional space, which can be presented in a two-dimensional

Table 8 The relevant relationships selected, with their partial correlation coefficient (r_p), the number of its degrees of freedom (df), and the level of the significance of the dependency given as a percentage (α). The names of variables are given in Appendix IV

Dependant	Determinant	r_p	df	α
SUCRAA	DENMAL	−0.26	93	1
	GUTCON	−0.40	93	1
SUCRAR	DENMAL	−0.25	80	5
	GUTCON	−0.38	80	1
SUCRRA	GUTCON	−0.15	66	>10
SUCRAE	DENEGG	−0.40	109	1
	GUTCON	−0.46	109	1
EFTIM	ACTMAL × DENMAL	−0.48	33	1
	DENWEB	0.36	33	5
	GUTCON	−0.72	33	1
EHTIM−EFTIM	EFTIM	0.32	32	5
	ACTMAL × DENMAL	−0.32	32	5
MAFTIM	GUTCON	−0.62	31	1
MAHTIM−MAFTIM	MAFTIM	0.69	31	1
	DENWEB	−0.56	31	1
COINAA	DENWEB	−0.63	6	10
	VELPRE	−0.64	6	10
COINAR	VELPRE	−0.66	6	10
COINAE	DENWEB	−0.53	4	>10
	DENMAL	−0.81	4	10
ACTMAL	DENMAL	−0.80	10	1
	DENWEB	0.92	10	1
ACTIM	DENMAL	0.46	59	1
	DENWEB	−0.43	59	1
RESTIM	ENCRRA × DISTUR	−0.47	47	1
	GUTCON	0.30	47	5
DISTUR	ACTMAL × DENMAL	−0.35	33	5
	GUTCON	0.37	33	5
VELMAL	DENMAL	−0.76	15	1
	DENWEB	0.53	15	5
	ACTMAL	−0.61	15	1
VELPRE	DENWEB	−0.57	8	10

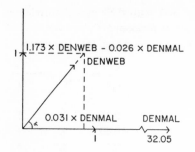

Fig. 9 | The orthonormalization of two vectors.

diagram. When XY represents the sum of products of the concomitant values of DENMAL and DENWEB and X^2 and Y^2 their sum of squares, then the cosine of the angle between the vectors is given by:

$$\cos \alpha = \frac{XY}{\sqrt{X^2 Y^2}}$$

In this example α equals $50°21'$. The lengths of the vectors are $\sqrt{X^2} = 32.05$ and $\sqrt{Y^2} = 1.108$. The purpose of the method is to define two vectors in the two-dimensional space, which have a unit length and an angle of 90° (correlation coefficient zero). These orthonormal vectors run into the direction of two perpendicular axes, which can be chosen arbitrarily in the two-dimensional space. The vectors have the common form $a \times$ DENMAL $+ b \times$ DENWEB. When such linear combinations are substituted in, for instance, the third degree polynomials of polyfactor analysis, the model of the multiple relationship of ACTMAL with DENMAL and DENWEB remains a third degree function, but becomes more descriptive due to the introduction of more terms (which is clear when the powers of the linear combinations are expanded). If there are sufficient data points, there is only one third degree function describing a model with the best fit to the values of ACTMAL measured. Polyfactor analysis approaches this model independently from the direction of the perpendicular axes chosen. Therefore, the most simple solution is chosen: according to Gram-Schmidt orthogonalization, one axis is taken into the direction of one of the vectors, say DENMAL, and the other is chosen perpendicular to the first. The first orthonormal vector has the length one in the direction of DENMAL: 0.031 DENMAL. The other vector is found as indicated in Fig. 9: 1.173 DENWEB − 0.026 DENMAL. These combinations will be termed orthonormal factors. The coefficients of all orthonormal factors derived are given in Table 9.

Also for the computation of orthonormal combinations there exists a mathematical method. The sums of products of concomitant values of two variables are computed for all pairs of rows and arranged in the matrix of inner products P. P is split into the product of a triangular matrix D and its transposed $(P = D'D)$ with a Choleski-procedure. The multiplication of D with the inverse of P gives a matrix with the coefficients of orthonormal combinations (DP^{-1}).

The values of the orthonormal factors were entered into the program for polyfactor analysis. The shapes of the curves obtained indicated, that they can be described satisfactorily by third degree polynomials $(aX^3 + bX^2 + cX + d)$. Because polynomials can be used conveniently to express the relationships in mathematical simulation models, polyfactor analysis was repeated, but with the subroutine that fits such polynomials by finding the regression coefficients $a - d$. The level of the dependants and the coefficients of the polynomials obtained are given in Table 9, where the coefficients d of the polynomials have been added to the level. The multiple relationships are defined by Table 9. So we have for ACTMAL:

$$ACTMAL = 0.71 - 3.19\ Y^3 + 3.69\ Y^2 - 1.08\ Y - 2.24\ Z^3 + \\ + 1.44\ Z^2 + 0.17\ Z$$

where

$$Y = 0.031\ DENMAL \text{ and } Z = -0.026\ DENMAL + 1.173\ DENWEB.$$

Also some conceivable upper or lower limits of the dependent variables are mentioned in Table 9. These limits will be used to prevent the computation of unreal, e.g. negative, values, when the determinants reach extreme values in simulation studies. It must be realized that polyfactor analysis, like all regression methods, gives approximate descriptions of the relationships only. Unreal results may occur, due to scarcity of data points. These have to be excluded in simulation.

Table 9 presents the ultimate results of the analysis of the continuous observations. If the prey densities, the density of the webbing cover, and the gut content of the predator are known at any one instant, likely estimates of the components of behaviour of the predator and the prey can be obtained one after another. The values of the ortho-

Table 9 The coefficients of the orthonormal factors ($co_1 - co_3$), the coefficients of the polynomia $aX^3 + bX^2 + cX$, the level (*lev*), and the lower and upper limits (l_l and l_u), as used for the quantitative description of the relevant multiple relationships in the mite system. The names of variables are given in Appendix IV (See opposite page)

Depend.	Determ.	co_1	co_2	co_3	a	b	c	lev	l_l	l_u
SUCRAA	DENMAL	0.013	0	0	-0.14	1.53	-1.27	0.15	0	1
	GUTCON	-0.010	0.231	0	11.30	0.59	-0.96			
SUCRAR	DENMAL	0.013	0	0	24.62	-9.62	-0.19	0.14	0	1
	GUTCON	-0.010	0.231	0	24.80	0.76	-1.51			
SUCRRA	GUTCON	0.223	0	0	-30.33	14.92	-2.08	0.09	0	1
SUCRAE	DENEGG	0.013	0	0	-45.83	23.85	-5.03	0.46	0.03	1
	GUTCON	-0.010	0.197	0	-27.56	-2.51	-0.59			
EFTIM	ACTMAL× DENMAL	0.042	0	0	-18.69	12.40	-1.51	0.10	0.01	—
	DENWEB	-0.036	0.648	0	0.88	0.77	-0.30			
	GUTCON	-0.008	-0.362	0.396	1.02	0.04	-0.26			
EHTIM-EFTIM	EFTIM	2.511	0	0	3.67	-1.61	0.20	0.005	0	—
	ACTMAL× DENMAL	-2.121	0.055	0	0.37	-0.16	-0.06			
MAFTIM	GUTCON	0.380	0	0	-27.82	15.60	-2.62	0.21	0.01	—
MAHTIM- MAFTIM	MAFTIM	1.114	0	0	-2.61	0.20	0.36	0.02	0	—

COINAR	VELPRE	-0.509	0.401	0	17.90	14.26	1.20	2.31	0	—
COINAE	VELPRE	0.383	0	0	-13.79	7.06	3.53	1.10	0	—
	DENWEB	0.958	0	0	-15.69	19.83	-6.79			
	DENMAL	-0.786	0.058	0	3.14	-3.67	0.68			
ACTMAL	DENMAL	0.031	0	0	-3.19	3.69	-1.08	0.71	0	0.9
	DENWEB	-0.026	1.173	0	-2.24	1.44	0.17			
ACTIM	DENMAL	0.011	0	0	9.30	-4.72	0.69	0.03	0.01	—
	DENWEB	-0.002	0.560	0	-1.73	-0.49	0.24			
RESTIM	ENCRRA×DISTUR	0.029	0	0	-3.32	1.93	-0.36	0.07	0.01	—
	GUTCON	-0.025	0.369	0	-3.84	0.69	0.18			
DISTUR	ACTMAL×DENMAL	0.032	0	0	-55.57	36.66	-5.91	0.66	0	1
	GUTCON	-0.026	0.393	0	9.20	-0.64	0.15			
VELMAL	DENMAL	0.024	0	0	17.03	0.02	-5.63	1.33	0.3	—
	DENWEB	-0.016	0.858	0	13.80	-1.72	-1.64			
	ACTMAL	-0.012	-0.973	0.616	-11.78	2.69	0.07			
VELPRE	DENWEB	0.953	0	0	-17.49	18.89	-6.09	1.36	0.3	—

Table 9 Caption, see opposite page

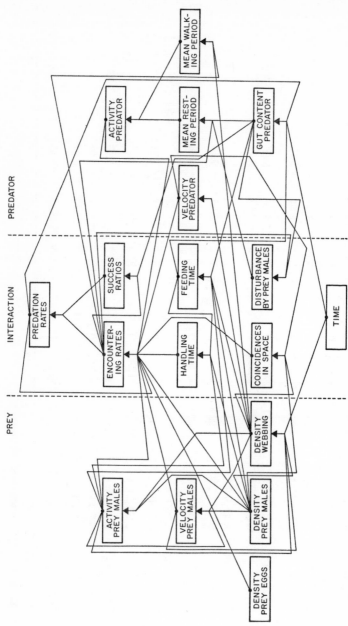

Fig. 10 | Conceptual model of a predator-prey system. One *Typhlodromus occidentalis* female on a leaf with constant numbers of *Tetranychus urticae* males and eggs.

44

normal linear combinations are computed first, and these are substituted in the sums of level and polynomials. In this way the relationships are represented rather cryptically, but also very concisely for incorporation into simulation models. By using these simulation models, it will be possible to judge the influence of each factor on the components of behaviour of the predator and the prey. This will be discussed in Chapter 5.

3.10 A conceptual model of the predator-prey system

Table 9 summarizes the relationships, which must be taken into account. Together with the encircled relationships in Table 1, they complete the conceptual model of the system, which is an imagination of the elements and relationships actually involved. This conceptual model is visualized in Fig. 10. In this diagram the elements are represented by blocks and the relationships by arrows, designating dependants and determinants. In the next chapter this conceptual model and the quantitative descriptions of the relationships will be used to build simulation models.

4 Simulation

The relationships derived in the foregoing chapters will now be used to build mathematical models of the system studied.

Some of the state variables considered, e.g. the gut content and the webbing density, vary with time, because they depend on processes like ingestion, digestion, and locomotion. Other system elements change because they are determined by state variables. The number of prey killed per time unit is also not constant. To study the dependency of predation on changes in predator and prey behaviour, as related to the densities of prey and webbing, models were used to simulate the dynamics of the system at different prey densities.

The method of simulation with mathematical models as applied to the analysis of ecological systems is discussed by Garfinkel (1965) and Patten (1971). For a comprehensive description of the general principles the reader is referred to this literature. The use of computers in the simulation of complex processes is essential because the integration of the changes of state variables is arithmetically laborious. Digital computers offer good facilities for programming and can be used for simulation when numerical integration methods are applied. Special programming languages for simulation with digital computers have been devised. The application of one of these languages, Continuous System Modeling Program (CSMP, IBM, 1968), in ecological modelling has been described by Brennan et al., (1970). The models derived in the following sections were programmed in CSMP, and the simulation was executed on the IBM 360/50 computer of the university of Nijmegen.

4.1 Updating

In the simulation programs using numerical integration, the state variables are termed integrals. The initial values of these integrals at time = zero are given and the rates of change are defined, for instance by differential equations. An updating routine computes new values of the integrals after a small time-interval D E L T from the old values and the rates of change. If a rectangular integration method is selected the rates of change as given by difference equations are used. For the next D E L T the rates, which are functions of the integrals, are recomputed and thus the cycle recurs until time reaches a preset level F I N T I M. This method of updating is principally different from the method used by Holling (1966) in the praying mantid-fly model (see Section 1.2).

D E L T may be fixed or variable. When variable it depends on the rates of change and on error tolerance. In the models derived in the following sections a fixed time-step is used, together with rectangular integration (Euler's method). Hence at each time-step the product of the rate of change and D E L T is added to the integrals. A small time-step (0.01 hours) was chosen, so as to make the probability of more than one capture in D E L T very close to zero, and to account for rapid processes like ingestion.

4.2   Stochastic versus deterministic simulation models

The purpose of our simulation models is the computation of the expectation values of the numbers of prey killed per predator, and other output variables, as a function of time. These expectation values are the average results of a large number of experiments with a single predator. Because the time between captures is a random variable with a certain probability distribution, the single experiments are stochastic processes with a variable number of captures within a limited time-interval. The expectation values are estimated by the total number of captures in a 'population' of experiments, divided by the number of experiments. The question is now, whether with the information obtained in the previous chapters a stochastic model is required to compute reliable expectation values, or whether a simple deterministic model will satisfy. The difference between these models is that a stochastic model accounts for the variance of stochastic system elements, while a deterministic model considers the expectation values of the variables only.

The philosophical backgrounds and the methodological consequences of stochastic versus deterministic models have often been discussed (e.g. Bailey, 1967; Kowal, 1971). The systematic differences between the methods, however, have never been explained comprehensively. This section gives a concise introduction to the theory of model building in this respect, and compares the different methods for simulating a simplified predation process.

Stochastic models simulate stochastic processes most realistically. In their simplest forms the values of stochastic variables are generated randomly according to their probability distribution in a series of consecutive time-intervals. The output of such models is a stochastic variable. The simulation is replicated, and the mean output value of replicates approaches the expectation value. This Monte-Carlo method is the original conception of simulation in the terminology of such mathematicians as Tocher (1963). It is, essentially, a sampling procedure.

Let Y be the number of prey killed by a predator in a simple predation process, and let the mean hourly velocity of locomotion V have a standard uniform probability distribution, with equal probabilities for equal intervals of values between zero and one. If the rate of change of Y is proportional to V:

$$d Y/dt = c V$$

then a stochastic model for a single experiment follows. To introduce the simulation language CSMP, the model will be stated as a CSMP program:

(a) *A stochastic model for a single process.* At every time-step DELT a value of V is generated with a random function generator. The rate of Y is computed, and the product of this rate and DELT is added to Y. When M is an arbitrary odd integer required to initiate the random function generator, YI is the initial value of Y at time zero, $c = 10$, DELT = 1 hour, the exposition time FINTIM = 24 hours, M = 3, and YI = 0, we have:

```
              V=RNDGEN(M)
   PARAMETER  M=3
   INCON      YI=0.0
   PARAMETER  C=10.0
              Y=INTGRL(YI,C*V)
   TIMER      DELT=1.0, FINTIM=24.0
   METHOD     RECT
```

This section of the program demonstrates the use of the INTGRL statement, which has two arguments separated by a comma: the initial value of the state variable, and its rate of change. The asterisk is the symbol for multiplication. The other statements show, how it is possible to declare the values of constant initial conditions, and time variables by using statement labels. The integration will be rectangular when RECT is used after the METHOD label.

The sum of Y (SUMY) of one thousand replicates of the experiment will be computed to obtain the expectation value of Y (EY) at the end of 24 hours. The replications of the simulation run are invoked in a TERMINAL section, which is not sorted by the CSMP compiler and can be written in FORTRAN. This program section also computes and prints EY. In FORTRAN it is programmed as follows:

```
TERMINAL
PARAMETER  NREP=1000.0
INCON      SUMY=0.0, TELLER=0.0
           SUMY=SUMY+Y
           TELLER=TELLER+1.0
           IF (TELLER.GE.NREP) GOTO 1
           CALL RERUN
           GOTO 2
      1    EY=SUMY/TELLER
           WRITE (6,100) EY
    100    FORMAT (H0,F10.4)
      2    CONTINUE
END
STOP
```

The TERMINAL section is executed only once in a run after TIME has reached FINTIM. The program is terminated by the two statements END and STOP. After END, new values of parameters can be given to repeat the simulation automatically. END marks the end of a run, while STOP indicates the end of the simulation.

In this example, $\mathscr{E}(Y)$ can be derived analytically since Y is a simple sum of stochastic variables:

$$\mathscr{E}(Y) = YI + 24\mathscr{E}(c\underline{V})$$

$$\mathscr{E}(c\underline{V}) = \int_0^1 cV \, dV = 5$$

$$\mathscr{E}(Y) = 0 + 24 \times 5 = 120$$

An introduction to the theory of the analytical evaluation of stochastic model output has been given by Bartlett (1960), and a general introduction to the theory of stochastic processes by Chiang (1968). Analytical solutions for complex models are not often available, because the inevitable integration is soon impracticable.

(b) *A stochastic model for a population.* In order to reduce the variance of the output variables, stochastic models may be applied to simulate the mean process in a large number of experiments with a single predator. Then characteristics of such a hypothetical population are considered. In the example used this may be the mean velocity in a number of experiments.

In a population of n single experiments the mean hourly velocity $\overline{V}$ has a probability distribution as well. W have:

$$\overline{V} = \sum_{1}^{n} V_i/n$$

where V_i represents the mutually independent velocities in the single experiments. According to the central limit theorem (Feller, 1968), $\overline{V}$ has a normal probability distribution when n is large. The expectation value and variance of V_i are known, because the velocities in the single experiments have a standard uniform probability distribution:

$$\mathscr{E}(V_i) = 1/2$$
$$\text{var}(V_i) = \int_0^1 V_i^2 \, dV_i - \left(\int_0^1 V_i \, dV_i \right)^2 = 1/12$$

Hence we have:

$$\mathscr{E}(\overline{V}) = n/n \times 1/2 = 1/2$$
$$\text{var}(\overline{V}) = n/n^2 \times 1/12 = 1/(12n)$$

The CSMP program for a stochastic model for a population will be identical to the program for model (a), but with the statement:

```
V=GAUSS(M,0.5,1.0/(12.0*N)**0.5)
```

instead of the statement

```
V=RNDGEN(M)
```

where N denotes the number of experiments. The GAUSS function generates random values from a normal distribution with the mean and standard deviation given as arguments.

The analytical solution is:

$$\mathscr{E}(\underline{Y}) = YI + 24\mathscr{E}(c\underline{V}) = 0 + 24 \times 5 = 120$$

For large n the variance of $\underline{V}$ approaches zero, and this model turns out to be deterministic for large populations. Then one run will be sufficient to compute a reliable estimate of $\mathscr{E}(\underline{Y})$.

(c) *A deterministic model for a population.* It is generally assumed that population processes can be adequately described by deterministic models, due to the phenomenon of variance reduction. This idea is inviting, because deterministic models require less computer time than stochastic models, are simpler to program, and simulation languages like CSMP offer better output facilities. In the example given, the CSMP program for a deterministic population model is very simple indeed:

Instead of the variable $\underline{V}$ its expectation value $\mathscr{E}(\underline{V}) = 0.5$ is used. We have:

```
PARAMETER  V=0.5,  C=10.0
INCON      YI=0.0
           Y=INTGRL(YI,C*V)
TIMER      DELT=1.0, PRDEL=1.0,
            FINTIM=24.0
METHOD     RECT
PRINT      Y
END
STOP
```

In this program the value of Y is printed at the end of each hour (PRDEL), together with the time. Analytically we have at the end of 24 hours:

$$Y = YI + 24c\mathscr{E}(\underline{V}) = 0 + 24 \times 5 = 120$$

(d) *A deterministic model for a single process.* Sometimes a deterministic model has been applied to a single process (e.g. Holling, 1966). A program for such a model in the example given is identical with the program of model (c). Analytically we have at the end of 24 hours:

$$Y = YI + 24c\mathscr{E}(\underline{V}) = 0 + 24 \times 5 = 120$$

Since the expectation value of the mean of a variable in a population with equal probability distribution is always equal to the expectation value of the variable itself, the models (c) and (d) are identical. In fact, type (d) is a representation of type (c), because individuals with consistently mean characteristics do not exist.

In spite of the theoretical differences, the models predict the same number of prey killed in 24 hour in the example employed. However, when the number of prey killed per hour is considered to be proportional to the square root of the hourly velocity $\underline{V}$, for example when the success ratio is proportional to $\underline{V}^{-.5}$, the situation is different. In this case the presumptions are:

$$d\underline{Y}/dt = c\underline{V}^{.5}, \text{ where c may be equal to 6}$$

$$\mathscr{E}(\underline{V}) = \mathscr{E}(\underline{\overline{V}}) = .5$$

$$\mathscr{E}(\underline{V}^{.5}) = \int_0^1 V^{.5}\,dV = .67$$

$$\mathscr{E}(\underline{\overline{V}}^{.5}) = \int_{-\infty}^{\infty} (6nV/\pi)^{.5}\,e^{-6n(V-.5)}\,dV \neq .67 \text{ for large n.}$$

The last equation can be derived from the density function of the normal probability distribution with mean $1/2$ and variance $1/(12n)$. The analytical solutions of the four models are:

(a) A stochastic model for a single process:

$$\mathscr{E}(\underline{Y}) = Y I + 24c\mathscr{E}(\underline{V}^{.5}) = 0 + 24 \times 4 = 96$$

(b) A stochastic model for a population:

$$\mathscr{E}(\underline{Y}) = Y I + 24c\mathscr{E}(\underline{\overline{V}}^{.5}) \neq 96$$

(c) A deterministic model for a population:

$$Y = Y I + 24c(\mathscr{E}(\underline{\overline{V}}))^{.5} = 144/2^{.5} = 101.8$$

(d) A deterministic model for a single process:

$$Y = Y I + 24c(\mathscr{E}(\underline{V}))^{.5} = 144/2^{.5} = 101.8$$

In this case the models (b), (c) and (d) are erroneous. Curvilinear functions of stochastic variables cause errors in simulation with deterministic models and models for populations. This is because the expectation value of curvilinear functions differs from the function of the expectation value.

Evidently, the choice between a model for a single process and for a population is prescribed by the type of observations and measurements. It was technically possible to observe the behaviour of individual predators in the mite system, but it would have been impracticable to measure concomitant mean values of behavioural components in a population. Therefore, the models (b) and (c) are not applicable to the mite system. For other systems it is sometimes more practicable to consider population variables, such as the mean length of age classes in the study of the production of fish populations (Beverton & Holt, 1957). In such cases a model for a population is indicated, and for large populations this may be deterministic. It has to be realized, however, that curvilinear relationships between stochastic properties of individuals and their contribution to the output of populations (e.g. the length of single fishes and the annual yield in weight of fish populations) may effect deviations of the results computed. The size of the deviations depends on the frequency distribution of the properties. Changes of the distribution may have considerable influence on the output of real populations, which is not expressed in the output of models for populations.

Sometimes the disproportion of the output of populations and the mean individual properties can be concealed in the values of population parameters. The relationship between the mean length of fishes and their age, for instance, may be established and defined for a certain population with a few parameters, e.g. with a von Bertalanffy growth equation (Beverton & Holt, 1957, p. 282). Other population parameters define the relationship between mean weight and mean length. In this way the model will be precise, but the objection to the change of the frequency distribution still holds. When the frequency distribution of the individual properties is variable, at least some of the population parameters are not constants.

Although they are less reliable than stochastic models, the simple deterministic population models are preferable for most purposes, when population parameters are available, and changing frequency distributions and curvilinear relationships do not prevail. Moreover, a stochastic model can only include the frequency distribution of a few variables, because the establishment of probability distributions requires much more a priori knowledge and experimentation than the determination of parameters. The accuracy gained by a stochastic approach may not be worthwhile in the analysis of complex systems.

In modelling the mite system, population parameters are not available and a choice has to be made between a deterministic and a stochastic model for a single process (type (a) or (d)). Both types, however, are unsuitable for large models with curvilinear relationships: type (a) is very expensive when used extensively, and type (d) is erroneous. Therefore, the main simulation study of the predation process has been performed with a new type of model, which has the advantages and disadvantages of both basic types, but the latter on an admissible level. Basically this is a multiple application of type (c) to classes of individuals. These classes are chosen in such a way that the crucial curvilinear functions are approximately linear within the classes. The principles of this method of compound simulation are discussed in the next section. Then models for stochastic, deterministic and compound simulation of the predation process of a single predator preying on eggs will be described.

4.3 Principles of compound simulation applied to classes of individuals

It is, in fact, possible to simulate a stochastic process with a stochastic model in one run, without the generation of random values. If the range of the standard uniform variable $\underline{V}$ in the example given in Section 4.2 is divided into s equal classes, the relative frequency of each class in an infinite number of single experiments is 1/s at each time-interval. If the number of prey killed per hour is $c\underline{V}^{.5}$, then the expectation value of the number killed ($\mathscr{E}(\underline{Y})$) increases with:

$$\mathrm{d}\mathscr{E}(\underline{Y})/\mathrm{d}t = \sum_{1}^{s} cV_{i}^{.5}/s$$

where V_i is the average value of $\underline{V}$ in the classes. When the classes are small, the function $\underline{V}^{.5}$ can be approached by a straight line within the classes (Fig. 11). When s = 1 the result of a deterministic model for a single process is obtained with little effort in computing; when s = ∞ we have the result of a stochastic model with much more effort. By weighing precision against expenses it is possible to find an optimum number of classes. Also the choice of the class boundaries is important (see Fig. 11).

In the process with $\mathrm{d}\underline{Y}/\mathrm{d}t = c\underline{V}^{.5}$ the behaviour of a predator depends on chance only. The sum of cV_{i}^{5}/s is constant. In simulation it is computed only once in an initial section prior to the computations repeated

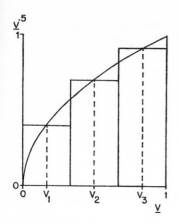

Fig. 11 | The partition of a standard uniform variable V into s classes with means V_i to reduce the difference between the sum of $V_i^{.5}/s$ and $_0\int^1 V^{.5}dV$.

in each time-interval. Such a Markov process is rarely met in real predator-prey systems. In general, the behaviour will depend on changing properties of the predator, which are state variables. In the example presented above, the success ratio may be assumed to be reciprocally proportional to ten times the gut content (G) plus one, so $d\underline{Y}/dt = c\underline{V}^{.5}/(10\,G+1)$. The change of the gut content depends on the number of captures in each time-interval, and hence on $\underline{V}$: e.g. $dG/dt = 3\,\underline{V}^{.5}/(10\,G+1)-0.5\,G$, where the first term represents ingestion, and the second term digestion. Because it is impossible to keep a record of the infinite number of individual gut contents, the range of possible G-values is divided into g classes with the actual average values G_j and the relative frequencies r_j. Now the increase of $\mathscr{E}(y)$ is given by:

$$d\mathscr{E}(\underline{Y})/dt = \sum_{j=1}^{g} r_j \sum_{i=1}^{s} cV_i^{.5}/(10\,G_j+1)/s \qquad (4.1)$$

The g×s subclasses are subsystems, which are all represented by a deterministic model; hence the name 'compound simulation'.

In this new example the values of V_i and s are constant, but those of G_j and r_j are variable and will be redefined at each time-interval. Each time-interval the g gut content classes split into s velocity classes with equal range, to be termed fractions, with a relative frequency r_j/s and mean velocity V_i. At the end of the time intervals all fractions have different gut contents, depending on G_j and V_i. The method reclasses the fractions in g gut content classes and computes the new values

55

G_j and r_j for the next time-interval. When the predator starts with an empty gut, the initial values of G_j and r_j are known.

To illustrate this cycle we may consider an example with three gut content classes (say, 0–0.3, 0.3–0.6, and 0.6–1.08 prey equivalents) and three velocity classes. Let, at the beginning of a time interval D E L T, the relative frequencies of the predators in the three classes be:

$r_j = 0.3, 0.1, 0.6$

with mean values of the gut content:

$G_j = 0.20, 0.40, 0.80$ prey equivalent.

In D E L T the three classes split into nine fractions with different velocities:

$V_i = 0.17, 0.5, 0.84, 0.17, 0.5, 0.84, 0.17, 0.5, 0.84$ m h^{-1}.

The values of V_i are the means of three classes with equal range between 0 and 1 m h^{-1}. The relative frequencies of the fractions equal one third of the relative frequencies of the gut content classes:

$r_i = 0.1, 0.1, 0.1, 0.03, 0.03, 0.03, 0.2, 0.2, 0.2$

At the end of D E L T the mean gut contents of the fractions are

$G_j + dG/dt \times$ D E L T.

When D E L T $= 1$ hour and

$dG/dt = 3 V_i^5/(10 G_j + 1) - 0.5 G_j,$

the new gut contents of the fractions are:

$G_i = 0.51, 0.81, 1.02, 0.44, 0.62, 0.76, 0.54, 0.64, 0.71$

The fractions are reclassed by comparing their mean gut content with the class boundaries. The fractions with values of G_i in italics go to the class 0.3–0.6 prey equiv., the others to the class 0.6–1.08 prey equiv. The new relative frequency of the classes is the sum of the relative frequencies of the contributing fractions. The new mean gut contents are the weighted means $(\Sigma r_i G_i)/(\Sigma r_i)$:

$r_j = 0, 0.33, 0.67$
$G_j = $ —, 0.52, 0.74 prey equivalents.

Thus, the relative frequency of individuals in a population is kept up

56

to date, and the rate of increase of the expectation value of the number of prey killed can be computed by Eqn (4.1).

There may be several state variables like the gut content. If they have a high variance in a population and are determinants in curvilinear relationships, these variables have to be divided into classes as well. the fraction values of all state variables have to be averaged, when the fractions are reclassed.

For general applications of this method a FORTRAN subroutine was developed to perform the reclassing of fractions. The routine is given in Appendix II. The application of this simulation method is illustrated in Section 4.4.3.

4.4 Simulation models

4.4.1 *Stochastic simulation of the predation on prey eggs*

A model of type (a) (see Section 4.2), using the Monte-Carlo technique, will be developed in this section to simulate a single predation process with one standardized *T. occidentalis* female preying on *T. urticae* eggs. This model was programmed to simulate one hundred replications of single experiments for different prey densities, to obtain the expectation values and the variances of the number of eggs destroyed, the biomass consumed, and the average gut content during the last six hours of a 24-h period.

The relevant system elements and their relationships have been discussed in Chapters 2 and 3. The symbolic names used there will be reused here. All names are defined in the context and are listed in Appendix IV. A diagram of the predation process is given in Fig. 12. A predator may start searching ($H = 0$, $S = 1$) at time zero. At the time t_1 it may catch a prey ($CATCH = 1$), to start a period of handling prey ($H = 1$, $S = 0$). The capture of a prey may take some time ($CATIM$). To account for this a state variable $RCTIM$ will be introduced. $RCTIM$ is set equal to $CATIM$ at the beginning of a handling period, and decreases with $DELT$ every time-interval until it is zero at the end of $CATIM$. At the end of $CATIM$ at t_2 the predator starts feeding. This increases the gut content ($GUTCON$), but decreases the content of the prey egg (EGG). At t_3 the prey egg may be empty, and, up to the next feeding period, $GUTCON$ will decrease by digestion. At t_4 the predator may abandon the prey ($ABAND = 1$) to start a new searching period.

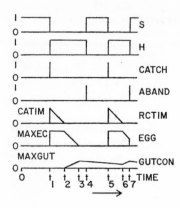

Fig. 12 | A time film of the predation process with a single predator.

At t_5 a second prey is seized, which may be abandoned before it is emptied.

The statements of a CSMP program simulating this process will be described in a sequence, which emphasizes the logical train of thought rather than the computational order. The instructions are sorted by the CSMP compiler, so the program can enter the computer in the same sequence.

The gut content of the predator. We may start with the gut content GUTCON, which is zero initially. Its rate of decrease by digestion (in the sense of the evacuation of food from the gut) is the product of GUTCON and the parameter b from Eqn (3.11). This parameter, denoted by DIGEST, was measured to be $0.435\ h^{-1}$ (see p. 24). The rate of increase by feeding is the ingestion rate INGRT. It follows:

```
GUTCON=INTGRL(0.0,-DIGEST*
    GUTCON+INGRT)
PARAMETER DIGEST=0.435
```

The ingestion rate depends on:
1 whether the predator is handling or searching prey (whether H equals 1 of 0),
2 whether it actually feeds on the prey handled (RCTIM equals 0),
3 on the contents of the prey egg (EGG greater than 0).

If IR is the ingestion rate during handling periods when the prey is not empty, EIRT is the ingestion rate when the predator actually

58

feeds on an egg, and REDHA is a reduction factor for the manipulation of prey during feeding periods, we may state:

$$IR=EIRT*REDHA*INSW(DELT-RCTIM,$$
$$0.0,1.0-RCTIM/DELT)$$
$$INGRT=H*INSW(EGG,0.0,1.0)*IR$$

The switching function INSW equals its second argument, if its first argument is negative; otherwise it is equal to its third argument.

For the ingestion rate EIRT the model represented by Eqn (3.1) is chosen. If MAXGUT is the maximum gut content M and INGEST is the parameter c from Section 3.3 when time is in hours, we state:

$$EIRT=INGEST*(MAXGUT-GUTCON)$$
$$PARAMETER \; INGEST=79.2, \; MAXGUT=1.08$$

Note, that the limited availability of food in a prey is accounted for by the definition of INGRT.

The engagement of the predator. The integer H is zero initially and becomes one, when the predator catches a prey (CATCH = 1). H becoms zero again, when the prey is abandoned. This is expressed by

$$FIXED \qquad H$$
$$H=0.5+INTGRL(0.0,$$
$$CATCH/DELT-ABAND/DELT)$$

Because in the conversion of real to integer values the compiler simply cuts the fraction, the addition of 0.5 prevents errors due to rounding off. Each time-interval, the rectangular integration method adds the product of CATCH/DELT-ABAND/DELT and DELT to H.

The time spent with handling prey (HTIM) has three components:
1 the time spent with recognizing and catching the prey (CATIM),
2 the time spent with feeding (FTIM),
3 the time spent with palpating and puncturing during the period of feeding.

As indicated by Table 8 the third component depends on the second. It may be proportionate, because the feeding is often interrupted for palpating and continued on other parts of the prey, especially when it is a male. In Fig. 13 the handling periods observed during continuous observation are plotted against the feeding time within these periods for eggs (no males present) and males. Both periods are in minutes.

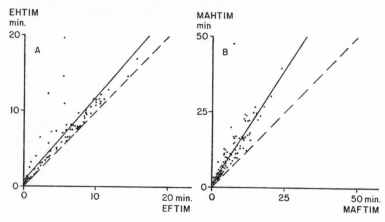

Fig. 13 | The length of periods of handling prey as related to the periods of feeding on eggs (A) and males (B).

It seems plausible to assume a linear relationship, which was determined by linear regression:

for eggs HTIM = 0.72 + 1.14 FTIM minutes,

for males HTIM = 0.84 + 1.53 FTIM minutes.

In these equations the intercepts with the ordinate represent CATIM, while REDHA is the reciprocal of the other coefficients (the slopes of the lines in Fig. 13). For eggs we have CATIM = 0.72/60 = 0.012 hours, and REDHA = 1/1.14 = 0.88, which is declared by the statement:

```
PARAMETER CATIM=0.012, REDHA=0.88
```

RCTIM becomes equal to CATIM when CATCH = 1. Then the next intervals it reduces with DELT, until it is lower than DELT. In the next time-interval it is set to zero, and it remains zero until the next capture. This can be expressed by:

```
RCTIM=INTGRL(0.0,RCTRT)
RCTRT=CATCH*CATIM/DELT-INSW
   (DELT-RCTIM,1.0,RCTIM/DELT)
```

The food content of the prey. The egg content EGG equals its maximum MAXEC when CATCH = 1, and it becomes zero when ABAND = 1.

It is reduced by feeding. MAXEC is given as L in Section 3.3, and has the value of 0.94 egg equivalents. We have:

```
        EGG=INTGRL(0.0,CATCH*
          MAXEC/DELT-ABAND*EGG/DELT-
          INGRT)
PARAMETER MAXEC=0.94
```

N.B. It is not likely that EGG becomes exactly zero during feeding periods. As soon as it becomes negative (only slightly), INGRT is set to zero.

The random variables. CATCH and ABAND are random variables, which may become one or zero during a time-interval. When RN is a random number between one and zero, and PRC and PRA are the probabilities that, respectively, a prey is caught during DELT when the predator is searching, and a prey is abandoned when the predator is handling prey, we have:

```
        S=1.0-H
        CATCH=INSW(PRC-RN,0.0,1.0)*S
        ABAND=INSW(PRA-RN,0.0,1.0)*H
        RN=RNDGEN(M)
INCON   M=3
```

The probability of a capture in DELT. Because the prey has a random spatial distribution, the number of prey encountered during a certain period t, $\underline{E}_t$, by a predator with a constant gut content will have a Poisson probability distribution:

$$P(\underline{E}_t = n) = \frac{(\bar{E}t)^n \, e^{-Et}}{n\,!}$$

where $\bar{E}$ is the average number of encounters per time unit. The probability of the number of prey caught during t, $\underline{C}_t$, is given by:

$$P(\underline{C}_t = n) = \sum_{m=n}^{\infty} P(\underline{E}_t = m) \times P(\underline{C}_t = n \mid \underline{E}_t = m)$$

where $P(\underline{C}_t = n | \underline{E}_t = m)$ is the conditional probability that the number of captures is n, if the number of encounters is m. This is given by the binomial probability distribution (Feller, 1968):

$$P(\underline{C}_t = n / \underline{E}_t = m) = \frac{m! \, s^n (1-s)^{m-n}}{n! \, (m-n)!}$$

where s is the success ratio.

We have:

$$\begin{aligned}
P(\underline{C}_t = n) &= \sum_{m=n}^{\infty} \frac{s^n (1-s)^{m-n} (\bar{E}t)^m \, e^{-Et}}{n! \, (m-n)!} \\
&= \frac{s^n (\bar{E}t)^n \, e^{-Et}}{n!} \sum_{m=0}^{\infty} \frac{(\bar{E}t - s\bar{E}t)^m}{m!} \\
&= \frac{s^n (\bar{E}t)^n \, e^{-Et}}{n!} \times e^{(Et - sEt)} \\
&= \frac{(s\bar{E}t)^n \, e^{-sEt}}{n!}
\end{aligned}$$

Consequently, irrespective of handling time, the number of prey caught in t also has a Poisson probability distribution, with average value $s\bar{E}t$. Let EC be the expected number of eggs caught in DELT, than the probability of zero captures in DELT is e^{-EC}. The probability of at least one capture is $1 - e^{-EC}$. Since only one capture is possible in DELT (because of the handling time), we may state:

$$PRC = 1.0 - EXP(-EC)$$

The expected number of prey caught per time unit is the product of the actual number of eggs on the leaf disk (ACNE), the coincidence in space (COINAE), the locomotion velocity of the predator (VELPRE), the activity of the predator (ACTPRE), and the success ratio (SUCRAE). If no males are present, COINAE and VELPRE are constants and are given by Table 9. Thus:

$$EC = ACNE * COINAE * VELPRE *$$
$$ACTPRE * SUCRAE * DELT$$
$$PARAMETER \ COINAE = 1.10, \ VELPRE = 1.36$$

The probability of abandonment in DELT. The abandonment of prey also depends on a Poisson process; during each very small instant of time there exists a very small probability that the predator leaves its prey. This probability seems to be determined by hunger in the first place, since hungry predators continue feeding on empty eggs (see

Section 3.3), while satiated predators abandon their prey even when it is not empty. Let EA be the average number of abandonments in DELT, if these are imaginarily conceived as repeatable events. Then:

$$PRA=1.0-EXP(-EA)$$

To find the relationship between EA and the gut content, the relative frequencies r_a of abandonments were computed for twenty equal gut-content classes, ranging from 0 to 1.08 egg equivalents, with an adapted version of ANCOWA, scanning the data cards of the continuous observations. When abandonments occurred in these classes, the relative frequencies were plotted for eggs (no males present) and males, see Fig. 14. The relative frequencies estimate the probability of an abandonment in the time periods T, during which the feeding predators remained in the different gut-content classes. The range of the classes is 0.054 egg equivalents, and T equals 0.054/IR. When IR is the ingestion rate during handling periods, we have:

$$r_a = 1 - e^{-(EA/DELT) \times (0.054/IR)} \text{ or:}$$
$$EA = -\ln(1-r_a)/ \, 0.054 \times IR \times DELT$$

IR is a function of the gut content, which has been defined for eggs on page 59.
The relationship between r_a and the gut content for feeding on eggs is satisfactorily described by a hyperbola (see Fig. 14):

$$r_a = \frac{p}{q-GUTCON} \text{ with } p \text{ and } q \text{ constants}$$

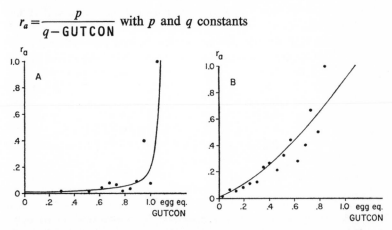

Fig. 14 | The frequency of abandonments as related to the gut content when the predator handles eggs (A) or males (B).

If it is assumed, that r_a equals one, when $GUTCON$ equals $MAXGUT = 1.08$, then we have:

$$r_a = \frac{p}{1.08 + p - GUTCON} \text{ or } 1/r_a = 1 + 1/p(1.08 - GUTCON)$$

Linear regression gives the value of p: $p = 0.013$. If $APAR1 = p$ and $APAR2 = p + 1.08$, it follows:

```
            EA=-ALOG(1.0-APAR1/(APAR2-
                GUTCON))/0.054*IR*DELT
PARAMETER APAR1=0.013, APAR2=1.093
```

The relationship between r_a and the gut content for feeding on males is not needed for this program, but will be used later. It is better described by a parabola than by a hyperbola:

$$r_a = p(GUTCON)^2 + q \times GUTCON + r, \text{ with } p, q \text{ and } r \text{ constants.}$$

Under the assumption, that r_a equals one, when $GUTCON$ equals 1.08, we have: $r = 1 - (1.08)^2 p - 1.08 q$. The parameters p and q were estimated by regression analysis and the descriptive model for males was found to be:

$$r_a = 0.47(GUTCON)^2 + 0.43 GUTCON + 0.009$$

The actual prey density. To maintain a constant prey density, the number of eggs NE on the leaf disk is replenished at the end of each half hour. The actual number of prey $ACNE$ equals NE initially. When $CATCH = 1$, $ACNE$ is reduced by one. At the end of the replenishment interval $REPDEL$, $ACNE$ is reset to NE by adding $REPL$. We have:

```
FIXED         NE,ACNE,REPL
              ACNE=0.5+INTGRL(NE,
              REPL/DELT-CATCH/DELT)
              REPL=IMPULS(0.0,REPDEL)*
              (NE-ACNE)
PARAMETER REPDEL=0.5
```

The $IMPULS$ function is one, when $TIME$ equals $REPDEL$ or a multiple of $REPDEL$, and zero otherwise. By changing this parameter it will be possible to study the influence of replenishment in experiments on predation. A sequence of values for NE can be given as a multiple

64

parameter. Then execution of the program is automatically repeated with the successive values of NE, so that compilation time is saved:

```
PARAMETER NE=(2,4,22,75,100)
```

The activity of the predator and the success ratio. The remaining variables ACTPRE and SUCRAE depend on the gut content and the density of the prey eggs DENEGG. The predator activity is determined by the mean walking period ACTIM and the mean resting period RESTIM:

$$ACTPRE=ACTIM/(ACTIM+RESTIM)$$

ACTIM and RESTIM are given by Table 9:

```
PARAMETER  ACTIM=0.03
           RESTIM=0.07+RTP
           RTP=(0.18+(0.69-3.84*RTF)*
           RTF)*RTF
           RTF=0.369*GUTCON
```

These statements demonstrate the use of the polynomials and the orthonormal coefficients described in Section 3.9. The success ratio SUCRAE is also given by Table 9. The ratio is multiplied by 0.25 (see Section 3.8). This may underestimate the ratio at high prey density (see Table 7). To avoid too low values, a lower limit of 0.0075 is applied. This value equals the success ratio at the density of 20 eggs/cm^2 in Kuchlein's experiments (Table 7) times 1/2. We have:

```
SUCRAE=AMAX1(0.03,0.46+
SRP1+SRP2)*0.25
```

This means that SUCRAE equals 0.0075, when 0.46 + SRP1 + SRP2 is lower than 0.03.

```
SRP1=(-5.03+(23.85-45.83*
SRF1)*SRF1)*SRF1
SRP2=(-0.59+(-2.51-27.56*
SRF2)*SRF2)*SRF2
SRF1=0.013*DENEGG
SRF2=0.197*GUTCON-0.010*DENEGG
DENEGG=ACNE/5.0
```

The output variables. The next statements compute the number of eggs captured (NPE), the biomass consumed (BIOMC), and the average gut content (AGUTC) during the last six hours of a 24-h period (to be identified by INT=1, otherwise INT=0):

```
NPE=INTGRL(0.0,CATCH/DELT*INT)
BIOMC=INTGRL(0.0,INGRT*INT)
INT=INSW(TIME-18.0,0.0,1.0)
AGUTC=INTGRL(0.0,GUTCON/DELT/
7.0*IMPULS(17.99,1.0))
```

AGUTC is the average value of the seven gut contents at the end of the last seven hours.

The time variables and the integration method are declared by:

```
TIMER      DELT=0.01, FINTIM=24.0
METHOD     RECT
```

A time-step of 0.01 hour will be sufficiently small to prevent the possibility of more than one capture in DELT, and to ensure a good account of ingestion.

In a TERMINAL section replications are invoked and the means and variances are computed and printed with the title 'TYPRES/ STOCHAST 1011711'. TYPRES denotes Typhlodromus Predation Process Simulator. The number is a date and sequence number for identification of the output. The section is not sorted, so the statements are given in the computational order. If NREP=100 is the number of replications, we have:

```
TERMINAL
PARAMETER  NREP=100
FIXED      NREP, TELLER
INCON      MNPE=0.0,MBIOM=0.0,
            MGUTC=0.0,SNPE=0.0,
            SBIOM=0.0,SGUTC=0.0
           MNPE=MNPE+NPE/NREP
           MBIOM=MBIOM+BIOMC/NREP
           MGUTC=MGUTC+AGUTC/NREP
           SNPE=SNPE+NPE**2
           SBIOM=SBIOM+BIOMC**2
           SGUTC=SGUTC+AGUTC**2
```

```
INCON          TELLER=0
               TELLER=TELLER+1
               IF (TELLER.GE.NREP) GOTO 1
               CALL RERUN
               GOTO 2
      1        VNPE=(SNPE-NREP*MNPE**2)/
                (NREP-1)
               VBIOM=(SBIOM-NREP*MBIOM**2)/
                (NREP-1)
               VGUTC=(SGUTC-NREP*MGUTC**2)/
                (NREP=1)
               WRITE (6,100) NE
    100        FORMAT(30H TYPRES/STOCHAST
                1011711 , NE= ,F8.0)
               WRITE (6,101)
    101        FORMAT(39H MNPE, VNPE, MBIOM,
                VBIOM, MGUTC, VGUTC)
               WRITE (6,102) MNPE,VNPE,
                MBIOM,VBIOM,MGUTC,VGUTC
    102        FORMAT(3(4X,2(F8.4,4X)))
```

The collective variables have to be initialized for the next series of reruns with a new NE value:

```
               TELLER=0
               MNPE=0.0
               MBIOM=0.0
               MGUTC=0.0
               SNPE=0.0
               SBIOM=0.0
               SGUTC=0.0
      2        CONTINUE
  END
  STOP
```

The predation rate (PRE) can be computed by dividing $MNPE$ by the number of hours of the exposition period concerned, which is six. The resulting mean values of the predation rate, the biomass consumed per hour and the gut content are plotted in Figs 16A, 20A, and 21A as functions of the egg density. The variances are used to derive the confidence intervals of the mean values in these figures. The execution time was about 28 minutes per prey density.

It must be realized, that this model is not completely stochastic. Only the most dominant stochastic variables are treated as such: the searching time and the handling time. Other variables, such as the feeding rate, are treated as determined variables, although they are certainly stochastic as well. Nothing is known, however, about their probability distribution, but it is assumed that their variance has little influence on the predation rate.

4.4.2 *Deterministic simulation of the predation on prey eggs*

The advantage of a deterministic model is its simplicity, the restriction to only one run, and the applicability of CSMP output facilities.

Only a few alterations are needed to obtain a deterministic version of the stochastic model described. H, S, CATCH, and ABAND represent conditions and events in the stochastic model; possible values are zero and one only. In a deterministic model these variables represent proportions on a continuous scale of individuals in a population, which in fact are subjected to these conditions or events. Also PRC and PRA are proportions now instead of probabilities, and all other variables represent population means. The complete program can be written as:

```
TITLE       TYPRES/DETERM 1011711
```

This title is printed as a heading above the printed results.

```
            GUTCON=INTGRL(0.0,-DIGEST*
            GUTCON+INGRT)
PARAMETER DIGEST=0.435
            IR=EIRT*REDHA*INSW
            (DELT-RCTIM,0.0,1.0-RCTIM/
            DELT)
            INGRT=H*INSW(EGG,0.0,1.0)*IR
            H=INTGRL(0.0,CATCH/DELT-
            ABAND/DELT)
            RCTIM=INTGRL(0.0,RCTRT)
            RCTRT=CATCH*CATIM/DELT-
            INSW(DELT-RCTIM,1.0,
```

```
              RCTIM/DELT)
PARAMETER  CATIM=0.012,  REDHA=0.88
           EGG=INTGRL(0.0,CATCH*MAXEC/
           DELT-ABAND*EGG/DELT-INGRT)
PARAMETER  MAXEC=0.94
           EIRT=INGEST*(MAXGUT-GUTCON)
PARAMETER  INGEST=79.2,  MAXGUT=1.08
           S=1.0-H
           CATCH=PRC*S
           ABAND=PRA*H
           PRC=1.0-EXP(-EC)
           PRA=1.0-EXP(-EA)
           EC=ACNE*COINAE*VELPRE*
           SUCRAE*ACTPRE*DELT
PARAMETER  COINAE=1.10,  VELPRE=1.36
           EA=-ALOG(1.0-APAR1/
           (APAR2-GUTCON))/0.054*IR*DELT
PARAMETER  APAR1=0.013,  APAR2=1.093
FIXED      NE,ACNE,REPL
           ACNE=INTGRL(NE,REPL/
           DELT-CATCH/DELT)
           REPL=IMPULS(0.0,REPDEL)*
           (NE-ACNE)
PARAMETER  REPDEL=0.5
PARAMETER  NE=(1,2,3,4,6,8,12,22,32,50,
           75,100)
           ACTPRE=ACTIM/(ACTIM+RESTIM)
PARAMETER  ACTIM=0.03
           RESTIM=0.07+RTP
           RTP=(0.18+(0.69-3.84*RTF)*
           RTF)*RTF
           RTF=0.369*GUTCON
           SUCRAE=AMAX1(0.03,0.46+SRP1+
           SRP2)*0.25
           SRP1=(-5.03+(23.85-45.83*SRF1)*
           SRF1)*SRF1
           SRP2=(-0.59+(-2.51-27.56*SRF2)*
           SRF2)*SRF2
           SRF1=0.013*DENEGG
```

```
              SRF2=0.197*GUTCON-0.10*DENEGG
              DENEGG=ACNE/5.0
              NPE=INTGRL(0.0,CATCH/DELT)
              BIOMC=INTGRL(0.0,INGRT)
TIMER         DELT=0.01, PRDEL=1.0,
               FINTIM=24.0
METHOD        RECT
PRINT         NPE, BIOMC, GUTCON
END
STOP
```

A comparison with the stochastic model shows that the terminal section and the random variable RN have disappeared, but that furthermore only a few structural statements have changed: the definitions of CATCH and ABAND. If PRDEL is declared 1.0, the values of TIME, NPE, BIOMC, and GUTCON are printed when TIME equals 0, 1, 2, 3,..., 24 hours. The program was executed, and PRE, BIOMRT and the average value of GUTCON in the last six hours have been plotted as a function of DENEGG in the Figs 16A, 20A, and 21A. The execution time was about 16 seconds for each prey density.

4.4.3 *Compound simulation of the predation process*

A model for compound simulation of the predation process outlined in Section 4.4.1 will be constructed in this section. Because emphasis is put on the modelling technique rather than on the description of the process, the reader must be prepared to go somewhat deeper into the topic of computer programming.

As mentioned in Section 4.3, the compound simulation method considers an infinite number of single processes, which are arranged into classes with respect to the values of their state variables. The predation process to be simulated takes place on a disk with a single predator and a certain density of prey eggs. So we will consider, say, a million disks with equal initial prey density. After a while, the state variables of these systems may have different values due to the stochastic character of the process.

The first matter of concern is the selection of classes. The values of stochastic state variables, which have a considerable variance and are determinants in curvilinear relationships, have to be divided into classes. In the process modelled, two state variables answer to this description: the gut content (GUTCON) and the engagement (H) of the predator. Therefore, we will consider five classes of GUTCON with the boundaries 0, 0.2, 0.4, 0.6, 0.8 and 1.08 egg equivalents, and two engagement classes: H = 0 and H = 1 with the fictive boundaries −0.5, 0.5 and 1.5. H = 0 indicates searching, H = 1 handling prey. A possible sequence of changes of the state of one predator is shown in Fig. 15.

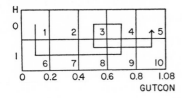

Fig. 15 | The arrangement of the state classes for compound simulation of the predation on eggs, with a record of the history of an individual predator.

The other state variables, the actual prey number (ACNE), the time needed to capture the prey (RCTIM) and the actual food content of the prey (EGG) do not have to be classified. ACNE is important for searching predators only, while RCTIM and EGG concern handling predators. ACNE and RCTIM do not vary widely, and EGG is only important in the few instances where it becomes zero. Little accuracy will be gained by dividing the values of these variables in classes.

At any time the million disks can be arranged in the ten gut content-engagement classes distinguished. In compound simulation the mean values of the state variables in each class are attached to every disk in the class, so disks in the same class are treated alike. This approach requires a special structure of the simulation program. But before we go into this, we will recapitulate what actually happens on the leaf disks.

At each time-interval DELT the five gut-content classes with searching predators each split into two fractions: one with a predator catching a prey, one with a predator continuing searching. The five classes with handling predators also split into two fractions: one with the predator abandoning the prey, the other with the predator continuing handling. Thus, the ten classes split into twenty fractions.

The structural statements applying to the classes with searching predators are derived in Section 4.4.1. They can be combined in a MACRO function to be used repeatedly when necessary. When used in a nonprocedural section of a CSMP program, a MACRO function writes these statements in the main program in the computational order (in the FORTRAN subroutine UPDATE, which is actually written and executed by a CSMP program). The following MACRO computes the relative frequency CATCH of disks with predators catching a prey in DELT, from the relative frequency of the class (RELF), the gut content (GUTCON), the actual prey number (ACNE), the coincidence in space (COINAE), and the velocity (VELPRE) and mean walking period (ACTIM) of the predator:

```
MACRO       CATCH=SEARCH(RELF,GUTCON,
             ACNE,COINAE,VELPRE,ACTIM)
            CATCH=(1.0-EXP(-EC))*RELF
            EC=ACNE*COINAE*VELPRE*
             SUCRAE*ACTPRE*DELT
            ACTPRE=ACTIM/(ACTIM+RESTIM)
            RESTIM=0.07+RTP
            RTP=(0.18+(0.69-3.84*RTF)*
             RTF)*RTF
            RTF=0.369*GUTCON
            SUCRAE=AMAX1(0.03,0.46+SRP1+
             SRP2)*0.25
            SRP1=(-5.03+(23.85-45.83*SRF1)*
             SRF1)*SRF1
            SRP2=(-0.59+(-2.51-27.56*SRF2)*
             SRF2)*SRF2
            SRF2=0.197*GUTCON-0.010*DENEGG
            SRF1=0.013*DENEGG
            DENEGG=ACNE/5.0
    ENDMAC
```

Another MACRO function computes the proportion ABAND of predators abandoning their prey, and the ingestion rate INGRT in the classes of handling predators, from RELF, GUTCON, RCTIM, EGG, REDHA, INGEST, MAXGUT, APAR1, and APAR2:

```
MACRO          ABAND,INGRT=HAND(RELF,GUTCON,
                 RCTIM,EGG,REDHA,
                 INGEST,MAXGUT,APAR1,APAR2)
               IR=EIRT*REDHA*INSW(DELT-
                 RCTIM,0.0,1.0-RCTIM/DELT)
               INGRT=INSW(EGG,0.0,1.0)*IR
               EIRT=INGEST*(MAXGUT-GUTCON)
               ABAND=(1.0-EXP(-EA))*RELF
               EA=-ALOG(1.0-APAR1/(APAR2-
                 GUTCON))/0.054*IR*DELT
ENDMAC
```

A title for the heading of program output and the parameters needed
are given by:

```
TITLE          TYPRES/COMPOUND 1011711
PARAMETER DIGEST=0.435
PARAMETER CATIM=0.012, REDHA=0.88
PARAMETER MAXEC=0.94
PARAMETER INGEST=79.2, MAXGUT=1.08
PARAMETER COINAE=1.10, VELPRE=1.36
PARAMETER REPDEL=0.5
PARAMETER APAR1=0.013, APAR2=1.093
PARAMETER NE=(1,2,3,4,6,8,12,22,32,50,
               75,100)
PARAMETER ACTIM=0.03
```

Now there remains only one problem: can the principles of compound simulation be applied in a CSMP simulation model? Here the
computer programming becomes rather sophisticated, and a good
understanding will require a basic knowledge of FORTRAN.

A program for compound simulation in fact repeatedly applies a deterministic model for all classes. Such a repetition of computations may
be invoked by using FORTRAN 'DO loops' (as explained below) in
procedural sections of a CSMP program. In FORTRAN the use of
indices and subscripted variables is very convenient. Therefore, the
mean values of class and fraction state variables will be stored in
2 two-dimensional matrices: $CLAV(i, n)$ and $FRAV(i, n)$. Then the
variable is indicated by the index i, and the class or fraction number
by the index n. The index $i = 1$ will be exclusively used for the relative
frequency. Thus we have:

the relative frequency	$RELF = CLAV(1, n)$
the gut content of the predator	$GUTCON = CLAV(2, n)$
the engagement of the predator	$H = CLAV(3, n)$
the actual number of prey	$ACNE = CLAV(4, n)$
the remnant of the period of catching	$RCTIM = CLAV(5, n)$
the prey content	$EGG = CLAV(6, n)$

In an initial section of the program the numbers of the gut content and engagement classes (NCL(1) and NCL(2)), the total number of classes (NCLASS) and fractions (NFRACT) are defined, and the initial values of the state variables are loaded into the CLAV matrix:

```
INITIAL
NOSORT
            NCL(1)=5
            NCL(2)=2
            NCLASS=NCL(1)*NCL(2)
            NFRACT=2*NCLASS
            DO 2 N=1, NCLASS
            DO 1 M=1,6
     1      CLAV(M,N)=0.0
     2      CLAV(4,N)=NE
            CLAV(1,1)=1.0
```

The INITIAL card precedes the statements, which have to be executed only once to define constants and initial conditions. The NOSORT card indicates that the statements in the section are given in the computational order and must not be sorted by the compiler. Sorting is not possible when the section contains FORTRAN control statements or subscripted variables left of the equal sign.

In a 'DO loop' all statements after the DO statement, until and including the statement with the number mentioned in the DO statement, are executed consecutively for a series of values of the running index named in the DO statement. This series comprises all the integers between and including the values indicated right of the equal sign.

Some variables used are integers, which has to be declared by:

```
FIXED        NCLASS,NFRACT,NCL,NE,N,N1,M
```

In the dynamic section the rates of change of the output variables NPE, BIOMC, and MGUTC (the mean gut content) are first set to zero. REPL is defined as an indicator of prey replenishment:

```
DYNAMIC
NOSORT
          NPERT=0.0
          BIOMRT=0.0
          MGUTRT=0.0
          REPL=IMPULS(0.0,REPDEL)
```

The values of the state variables in the different fractions (FRAV(i, n))
are determined, and the rates of change of the output variables are
accumulated in a DO loop. The index N refers to the class considered
and to the fraction with a predator continuing its engagement, N1
to the fraction with a predator changing its engagement. For the sake
of clearness the state variables of the classes are renamed with their
own symbolic name. They consequently have two names:

```
          DO 6 N=1,NCLASS
          N1=N+NCLASS
          RELF=CLAV(1,N)
          IF (RELF.LT.ZERO) GOTO 5
CONSTANT  ZERO=1.E-05
```

The program jumps to statement 5 when RELF is zero to avoid un-
necessary computations for empty classes. This also prevents underflow
in emptying classes during simulation.

```
          GUTCON=AMIN1(1.08,CLAV(2,N)*
          INSW(CLAV(2,N),0.0,1.0))
```

This statement keeps GUTCON within its limits in spite of rounding
off and integration errors.

```
          ACNE=CLAV(4,N)+
          REPL*(NE-CLAV(4,N))
          RCTIM=CLAV(5,N)
          EGG=CLAV(6,N)
```

The classes with searching predators are considered first. They split
into two fractions. The relative frequency of disks with a predator
catching a prey is computed:

```
                    IF (N.GT.5) GOTO 4
       SORT
                    CATCH=SEARCH(RELF,GUTCON,ACNE,
                       COINAE,VELPRE,ACTIM)
       NOSORT
```

The relative frequencies of the two fractions are:

```
              FRAV(1,N)=RELF-CATCH
              FRAV(1,N1)=CATCH
```

The gut content of both fractions decreases by digestion:

```
              FRAV(2,N)=GUTCON-
               GUTCON*DIGEST*DELT
              FRAV(2,N1)=FRAV(2,N)
```

The engagement of the predator is indicated by:

```
              FRAV(3,N)=0.0
              FRAV(3,N1)=1.0
```

The actual number of prey reduces by one in the fraction with a predator catching a prey, but cannot become negative:

```
              FRAV(4,N)=ACNE
              FRAV(4,N1)=AMAX1(0.0,ACNE-1.0)
```

The capture of a prey requires some time:

```
              FRAV(5,N)=0.0
              FRAV(5,N1)=CATIM
```

When a prey is captured, its food content is maximum:

```
              FRAV(6,N)=0.0
              FRAV(6,N1)=MAXEC
```

The expected number of prey killed increases in DELT with the sum of CATCH in all classes with searching predators:

```
              NPERT=NPERT+CATCH/DELT
```

The mean gut content decreases by digestion:

```
              MGUTRT=MGUTRT-RELF*DIGEST*GUTCON
              GOTO 6
```

Then the classes of handling predators are considered. They also split into two fractions. The relative frequency of disks with a predator abandoning its prey, and the ingestion rate, are computed:

```
     4          CONTINUE
SORT
                ABAND,INGRT=HAND(RELF,
                GUTCON,RCTIM,EGG,REDHA,
                 INGEST,MAXGUT,APAR1,APAR2)
NOSORT
```

The state variables of the fractions are defined as described above for the classes with searching predators:

```
                FRAV(1,N)=RELF-ABAND
                FRAV(1,N1)=ABAND
                FRAV(2,N)=GUTCON-GUTCON*DIGEST*
                 DELT+INGRT*DELT
                FRAV(2,N1)=FRAV(2,N)
                FRAV(3,N)=1.0
                FRAV(3,N1)=0.0
                FRAV(4,N)=ACNE
                FRAV(4,N1)=ACNE
                FRAV(5,N)=RCTIM-INSW(DELT-
                 RCTIM,DELT,RCTIM)
                FRAV(5,N1)=0.0
                FRAV(6,N)=AMAX1(0.0,EGG-
                 INGRT*DELT)
                FRAV(6,N1)=0.0
```

Both the biomass consumed and mean gut content increase by ingestion, the latter decreases by digestion:

```
                BIOMRT=BIOMRT+RELF*INGRT
                MGUTRT=MGUTRT+RELF*(INGRT-
                 DIGEST*GUTCON)
                GOTO 6
```

For empty classes we state:

```
     5          FRAV(1,N)=0.0
                FRAV(1,N1)=0.0
```

The loop is terminated by:

```
6        CONTINUE
```

After completion of the loop the fractions have to be reclassed. For general applications of this method a FORTRAN subroutine, named RECLAS, has been developed to perform the necessary computations. This subroutine requires as input arguments a previously defined matrix of fraction variables (FVAR), an array of class boundaries (BO), the total number of fractions (NFT), an array with the numbers of classes of the variables divided in classes (NC), the number of state variables divided in classes (NCV), and the total number of state variables (NVT). Its output argument is a matrix of class variables (CVAR). The subroutine is given in Appendix II. Its general form is:

```
SUBROUTINE RECLAS(FVAR,BO,NFT,
NC,NCV,NVT,CVAR)
```

In the present simulation program we state:

```
CALL RECLAS(FRAV,BOUND,NFRACT,
NCL,2,6,CLAV)
```

For general purposes the dimensioned variables FVAR, CVAR, BO and NC have to be given ample space in the memory of the computer. They are dimensioned to allow for nine state variables, three variables divided in maximally five classes, fifteen classes and thirty-five subclasses. The actual arguments have to be dimensioned accordingly in the main program calling the subroutine. Hence, space in the computer is reserved by:

```
/          DIMENSION CLAV(9,15), FRAV(9,35)
STORAGE    NCL(3),BOUND(18)
```

The class numbers are given in the initial section. The class boundaries are given by:

```
TABLE      BOUND(1-6)=.0,.2,.4,.6,.8,1.08,
           BOUND(7-9)=-.5,.5,1.5
```

The output variables are defined by:

```
NPE=INTGRL(0.0,NPERT)
BIOMC=INTGRL(0.0,BIOMRT)
MGUTC=INTGRL(0.0,MGUTRT)
```

The program is terminated by:

```
TIMER          DELT=0.01, PRDEL=1.0,
               FINTIM=24.0
METHOD         RECT
PRINT          NPE,BIOMC, MGUTC
END
STOP
```

The subroutine RECLAS has to be added to the program after the STOP statement.

Compound simulation was used in a comprehensive model for predation on males and eggs exposed simultaneously. The presence of males means that the model must be substantially extended as shown in the relational diagram in Fig. 10. New elements, like the webbing density and disturbance, have to be incorporated. Nevertheless, the main principles applied in the model for eggs remain unaltered. Essentially, it has the same program, but the handling predators are distinguished in two classes: those handling a prey egg, and those handling a male. The statements subjoined are evident from the relationships quantified in Chapter 3. The names of the variables used are listed in Appendix IV. The program is given in Appendix III.

The program for eggs and males was executed for various values of the number of eggs and males present, the replenishment interval, and the initial distance covered by the males. The results have been plotted in Figs 16 to 29. They include the numbers of prey captured and the biomass consumed per hour, the mean gut content, the mean actual prey densities, prey mortality rate and utility, the density of webbing, and the proportion of time the predator spent handling. The execution time was about fourteen minutes per run on the IBM 360/50. The much simpler stochastic model in Section 4.4.1 for eggs only, consumes twice as much time already. The use of a stochastic version of the model for eggs and males together would be very expensive, unless a low number of replicate runs is executed, which lowers accuracy.

4.5 Simulated responses to prey density

The system studied has several decision variables: the numbers of prey maintained on the leaf disk, the prey replenishment interval, the initial density of the webbing cover as determined by the initial distance covered by the prey males (IDIST), and the initial gut content of the predator. By changing parameter cards the models derived are easily adapted to optional combinations of these variables. In this way it is possible to simulate the dynamics of the system during a 24-h period for each situation, and to compare the results.

According to Kuchlein's conditions the replenishment interval was set to 0.5 hours, the initial distance covered to 0.01 m, and the initial gut content (of a 24-h period) to 0.0 egg equivalents. The prey densities used by him were also accepted: different numbers of eggs or males, or a combination of eighty eggs and a variable number of males. To evaluate the influence of prey replenishment and the webbing density, the program for the compound simulation of the predation on males and eggs (COMPOUND) was executed for some of the prey densities with REPDEL = 50 hours, IDIST = 0.01 m, FINTIM = 6.0 hours and REPDEL = 0.5 hours, IDIST = 100.0 m, FINTIM = 6.0 hours. The intrinsic rate of prey mortality (see Section 5.2.6), computed as the proportion of ACN killed in DELT, and the mean gut content are plotted in Fig. 22 as a function of the time for three densities of eggs and males. The relationships indicate that these values are not constant during the 24-h period. Therefore, a distinction has been made between the values of output variables in the last and the first six hours of the 24-h period.

The resulting values of the gut content and the density of the webbing cover were averaged for the six hour time-intervals. Together with the prey densities, these average values were entered into the relationships recorded by Table 9. In this way the mean values of all behavioural components could approximately be derived for the periods concerned. To perform the necessary computations a program EXPLAIN was constructed. The values, which correspond with the results of the program COMPOUND, are given in the Tables 10–14.

5 Discussion

5.1 The functional response to prey density

Figures 16–18 depict the relationships between the numbers of prey killed per hour and the prey density as generated by the models under the conditions indicated. None of these functional response curves correspond exactly with the fundamental types distinguished by Holling (1959a, 1961). Holling's Type 2 response for invertebrates, a gradually levelling curve, bears the closest resemblance. However, for predation of eggs we find a domed curve and the levelling is less gradual. The decline at high prey density is mentioned by Holling as an effect of swarming by prey or stimulus satiation. For predation of males we find an increasing slope at high prey density, and with both eggs and males present the curves are much more complicated than Holling's types.

Functional response curves deviating from the fundamental types have frequently been observed in experiments with acarine or arachnid predators (Chant, 1961b; Haynes & Sisojevic, 1966; Mori, 1969; Mori & Chant, 1966; Sandness & McMurtry, 1970). The explanations of the deviations given are mainly based on assumptions, because the

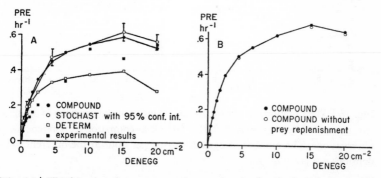

Fig. 16 | The functional response to the egg density for the last six hours (A) and the first six hours (B) of a 24-h period.

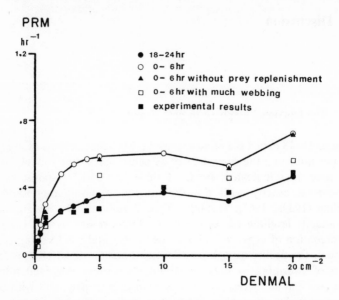

Fig. 17 | The functional response to male density as computed by the program COMPOUND, and obtained by independent experiments.

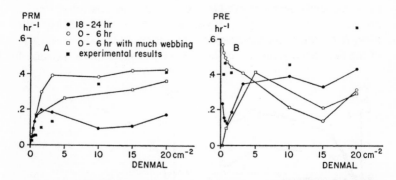

Fig. 18 | The functional response to male density in the presence of 16 eggs/cm^2 computed by the program COMPOUND as compared with independent observations. The number of eggs destroyed per hour at the different male densities are given in B.

Table 10 Mean values of behavioural components at different densities of prey eggs during the period 18–24 hour

DENEGG	0.2	0.4	0.6	0.8	1.2	1.6	2.4	4.4	6.4	10.0	15.0	20.0
GUTCON	0.10	0.16	0.21	0.25	0.31	0.36	0.41	0.48	0.51	0.54	0.58	0.54
RESTIM	0.06	0.06	0.07	0.07	0.08	0.08	0.08	0.08	0.08	0.08	0.08	0.08
ACTIM	0.03	0.03	0.03	0.03	0.03	0.03	0.03	0.03	0.03	0.03	0.03	0.03
VELPRE	1.36	1.36	1.36	1.36	1.36	1.36	1.36	1.36	1.36	1 36	1.36	1.36
COINAE	1.10	1.10	1.10	1.10	1.10	1.10	1.10	1.10	1.10	1.10	1.10	1.10
SUCRAE	.110	.105	.101	.096	.089	.082	.071	.052	.040	.028	.018	.008
EFTIM	0.09	0.09	0.08	0.08	0.07	0.07	0.06	0.06	0.06	0.06	0.06	0.06
EHTIM	0.11	0.10	0.09	0.09	0.08	0.08	0.08	0.08	0.07	0.07	0.07	0.07

83

curves have not been generated by mathematically defined models. The models derived in the foregoing chapter also produce deviating responses, but these can be explained by the changes of the system elements involved. In this section the functional response curves obtained by the program COMPOUND will be compared with the experimental results of Kuchlein. The factors responsible for their shape will be discussed briefly. They will be dealt with more extensively in Section 5.2.

5.1.1 *The predation of eggs*

The general shape of the functional response to egg density shows a decreasing slope (Fig. 16) like the Type 2 response described by Holling. It is generally believed that such a levelling can be induced by an increasing proportion of handling time, as it is formally described by Holling's disk equation (Holling, 1959b). The predator's occupation with handling prey, however, hardly exceeds 2% of the exposition time even at the highest prey densities (Fig. 19). As far as searching time is concerned, the number of prey killed per hour must almost be proportional to the prey density, so there must be other factors responsible for the decreasing slope. From Table 10 it can be concluded that the slope is mainly caused by a decrease of the success ratio, and to a lesser extent by an increase of the mean resting period.

The main determinant of the shape of the functional response curves is the decrease of the success ratio with increasing prey density, which follows from the negative correlation of the success ratio with the gut

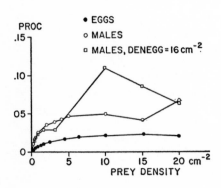

Fig. 19 | The proportion of time the predator is occupied with handling prey in the period 18–24 hour as related to prey density. Computed by the program COMPOUND.

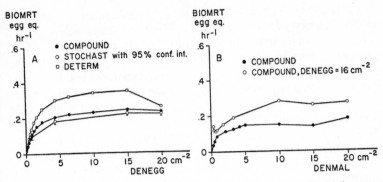

Fig. 20 | The biomass consumed per hour in the period 18–24 hour as related to egg density (A) and male density (B).

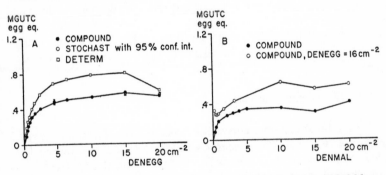

Fig. 21 | The mean gut content of the predator during the period 18–24 hour as related to egg density (A) and male density (B).

content (see Table 8). This is rather surprising, because for invertebrate predators such an important effect of hunger has only been described in connection with a variable distance of reaction (Holling, 1966). *T. occidentalis* behaves more like the stickleback (Beukema, 1968). The curves obtained by simulation are not completely identical to the fundamental Type 2 response. There seems to be an inflexion point at about four eggs per cm², after which the curves rise more linearly. The reduction of the success ratio decreases between four and fifteen eggs per cm², probably due to a lower increase of the gut content in this range (Fig. 21A).

The decrease of the predation rate at high egg densities is caused by a sudden decrease of the success ratio. This ratio is not only negatively correlated with the gut content, but also with the egg density (Table 8). The last correlation, which is also found for males, has to be attributed to a direct influence of encounters on the success ratio, e.g. by a reduction of the responsiveness of the predator by stimulus satiation at a high frequency of encounters. This effect will be termed 'inhibition by prey' according to Holling (1966).

As mentioned in the introduction and in Section 2.1, we can compare the theoretical curves with the independent observations of Kuchlein. For eggs, the confidence interval of these observations is given in Fig. 2A and the mean values are plotted in Fig. 16A. The wide confidence intervals and the discrepancies induced by the difference in observation methods (see Section 3.8) make a comparison of the details of the curves too inconclusive. Nevertheless, the decreasing slope and the inflexion point are distinct in the empirical results, and these can now be explained. The abrupt levelling in Kuchlein's observations at high prey densities may be related to the decline in the theoretical curves.

5.1.2 *The predation of males*

The functional response curves for males are also basically a Type 2 response (Fig. 17). The predator's occupation does not exceed 10% (Fig. 19 for males) and follows the predation rate closely. Hence, it is not an important determinant of the predation rate, otherwise they would be inversily correlated. For an explanation of the intricate curve we consider the behavioural components in Table 11.

Four properties of the curve computed for 18–24 hour, which are also reflected by Kuchlein's observations, have to be explained:

1 A general rise to a level that is reached at the density 10 males/cm^2
2 A superimposed increasing slope between the densities 2 and 5 males/cm^2
3 A decline between the densities 10 and 15 males/cm^2
4 An upward inclination beyond the density 15 males/cm^2.

The main cause of levelling is again the decrease of the success ratios with increasing prey density. The rise to a level is less gradual for males, than it is for eggs. This is explained by Table 11, which shows that in addition to the success ratios also the locomotion velocity of the

Table 11 Mean values of behavioural components at different densities ot prey males during the period 18–24 hour

DENMAL	0.2	0.4	0.8	2.0	3.0	4.0	5.0	10.0	15.0	20.0
GUTCON	0.09	0.14	0.21	0.27	0.30	0.32	0.34	0.36	0.31	0.42
DENWEB	0.17	0.24	0.31	0.38	0.40	0.41	0.41	0.43	0.47	0.57
RESTIM	0.05	0.04	0.03	0.03	0.03	0.03	0.03	0.02	0.06	0.01
ACTIM	0.05	0.05	0.06	0.06	0.07	0.07	0.08	0.08	0.07	0.09
VELPRE	0.79	0.75	0.76	0.81	0.82	0.83	0.83	0.84	0.86	0.83
ACTMAL	0.77	0.78	0.78	0.75	0.74	0.72	0.72	0.73	0.72	0.57
VELMAL	1.04	1.02	0.98	0.82	0.69	0.55	0.42	0.30	0.30	1.60
DISTUR	0.64	0.61	0.57	0.46	0.41	0.39	0.39	0.56	0.63	0.63
COINAA	3.30	2.81	2.19	1.18	0.82	0.70	0.74	0.27	0.00	0.00
COINAR	3.64	3.57	3.59	3.66	3.68	3.69	3.68	3.71	3.73	3.70
COINRA	1.58	1.58	1.58	1.58	1.58	1.58	1.58	1.58	1.58	1.58
SUCRAA	0.032	0.029	0.026	0.020	0.017	0.015	0.012	0.007	0.008	0.003
SUCRAR	0.029	0.025	0.021	0.018	0.017	0.015	0.013	0.007	0.006	0.001
SUCRRA	0.015	0.011	0.007	0.004	0.003	0.003	0.002	0.002	0.003	0.001
MAFTIM	0.14	0.11	0.09	0.08	0.08	0.07	0.07	0.08	0.07	0.08
MAHTIM	0.22	0.18	0.14	0.12	0.11	0.11	0.11	0.11	0.11	0.11

males, and the coincidence in space of the walking predator and walking males decrease with increasing male density. The increase in activity of the predator counteracts the levelling to some extent.

The upward incline between the densities 2 and 5 males per cm^2 is explained by the coincidence in space of walking predator and walking males (COINAA). This variable is rather constant in this range, but decreases rapidly at other male densities. This demonstrates that an intermission of a levelling effect in some prey density traject can cause an indentation of the functional response curve. The reduction of COINAA is related to the increasing density of webbing and increasing aggregation of the prey males. The first factor operates mainly in the lower density range, the latter at high prey densities. This can explain, why at moderate male densities there is little increase of the effect. The cause of the reduction of COINAA is obvious: the predator is unable to walk on the webbing cover as done by the males, and it is often frightened by the dense clusters of males at high male density. Predation rate reducing influences of webbing and aggregation were also observed by Mori (1969), when he studied the predation on *T. urticae* female adults and deutenomphs by *Amblyseius longispinosus*. Mori obtained domed functional response curves, but his highest prey density was 5.12 prey per cm^2.

The spectacular increase of the predation rate beyond 15 males/cm^2 is brought about by an increase of predator activity by disturbance and a decrease of male activity. Both increase the rate of encounters, because the coincidence in space with resting males remains high. Presumably, the negative correlation of the male activity with the male density (see Table 8) overrides the positive correlation with the webbing density. While webbing facilitates the male locomotion and separates the males from their feeding sites, the presence of other males seems to provide tactile or olfactory stimuli reducing locomotion.

The conformity of the theoretical curve for the 18–24 hour period and the results of Kuchlein's observations is surprisingly good. This justifies the conclusion that factors such as mentioned above can give an unimaginable shape to the functional response in real systems, which is difficult to interpret without a sufficiently detailed model.

5.1.3   *The predation of eggs and males together*

As indicated in Table 12, the slope of the curve for males in the presence

Table 12 Mean values of behavioural components at different densities of prey males in the presence of 16 prey eggs per cm² during the period 18–24 hour

	0.2	0.4	0.6	0.8	1.6	3.2	10.0	15.0	20.0
DENMAL	0.33	0.29	0.28	0.29	0.35	0.44	0.64	0.57	0.62
GUTCON	0.18	0.24	0.28	0.30	0.37	0.41	0.43	0.47	0.56
DENWEB									
RESTIM	0.06	0.05	0.05	0.04	0.03	0.04	0.03	0.01	0.01
ACTIM	0.05	0.05	0.06	0.06	0.06	0.07	0.08	0.07	0.09
VELPRE	0.79	0.75	0.75	0.76	0.80	0.82	0.84	0.86	0.83
ACTMAL	0.77	0.78	0.78	0.78	0.76	0.73	0.73	0.72	0.57
VELMAL	1.04	1.02	1.00	0.97	0.88	0.66	0.30	0.30	1.60
DISTUR	0.66	0.62	0.60	0.57	0.50	0.42	0.58	0.69	0.67
COINAA	3.29	2.81	2.47	2.24	1.43	0.80	0.29	0.00	0.00
COINAR	3.64	3.57	3.57	3.59	3.64	3.68	3.71	3.73	3.70
COINRA	1.58	1.58	1.58	1.58	1.58	1.58	1.58	1.58	1.58
COINAE	0.30	0.18	0.14	0.14	0.19	0.37	0.57	0.45	0.51
SUCRAA	0.021	0.023	0.022	0.021	0.017	0.011	0.000	0.000	0.000
SUCRAR	0.012	0.014	0.015	0.015	0.012	0.007	0.000	0.000	0.000
SUCRRA	0.002	0.004	0.004	0.003	0.002	0.001	0.003	0.002	0.003
SUCRAE	0.024	0.026	0.026	0.025	0.023	0.020	0.015	0.017	0.015
MAFTIM	0.07	0.08	0.08	0.08	0.07	0.08	0.10	0.09	0.10
MAHTIM	0.12	0.12	0.12	0.12	0.11	0.12	0.14	0.14	0.14
EFTIM	0.05	0.05	0.05	0.05	0.03	0.03	0.25	0.25	0.19
EHTIM	0.07	0.07	0.07	0.06	0.05	0.04	0.64	0.67	0.29

of eggs is determined mainly by the decreasing success ratios and the decreasing coincidence in space of active animals. The success ratios SUCRAA and SUCRAR almost reach zero at 10 males/cm^2, producing a steep decline in predation rate. This is reinforced by an increased predator occupation (Fig. 19), due to an increased feeding time for eggs. The latter may be somewhat exaggerated due to scarcity of data points, but the shelter provided by webbing can induce prolonged feeding on eggs.

At high male densities, the male predation rate increases again by the relative constancy of the components involved and by an increase of the male locomotion velocity. At these densities most males were killed, because they bumped into the resting predator.

In the presence of sixteen eggs per cm^2 the functional response curve for males for the period 18–24 hour differs widely from the experimental results (Fig. 18A). This may be due to a high webbing density or a too short adaptation period in Kuchlein's experiments. Both may be true, because the eggs on the disks used in his experiments were laid by female spider mites, which remained on the disks for a sufficiently long time to produce a high density of eggs. Afterwards, the females and the surplus eggs were removed from the disks, but the females presumably produced much webbing. Long adaptation periods were not applied in combination with this method of disk preparation. Kuchlein's results suggest that at high male density and in the presence of sixteen eggs per cm^2 the predator kills almost as many eggs and males as two predators would do together, one at the same density of males, and the other at the same density of eggs (compare Fig. 18 with Figs 16A and 17). Since the feeding on each kind of prey increases the gut content, and hence decreases the success ratio with regard to alternative kinds of prey, this result seems to be unrealistic. Probably, both the webbing density and the short adaptation period have played a role.

Although the prey eggs are always given in the same density, the predation rate of eggs in the polyphagous situation is not constant (Fig. 18B). The agreement with experimental results is best for the final 6-h interval. The theoretical and experimental results both indicate, that the predation rate for eggs increases with the male density, a fact which at first seems unrealistic. Table 12 shows that this increase is caused by an increase in the predator's activity and velocity, and by an increase of the coincidence in space. These factors overrule the influence of the gut content on the success ratio. The predator's activity is increased by the disturbance by males, so that the mean resting period is shortened and the mean walking period lengthened. The coincidence in space and the predator's velocity are greatly reduced at moderate densities of males and webbing, but are partly restored if both increase. This may be because the predator is gradually restricted to areas not preferred by the males.

5.2 The influence of some factors on the functional response

Although there is some doubt concerning the reliability of the relationships gained by polyfactor analysis, all phenomena influencing the functional response to prey density (to be termed factors) are conceivably real. Simulation makes clear that such factors can easily make functional response curves unimaginable and hardly interpretable. Hence Holling's idea that the shape of functional response curves can be classified in a few fundamental types is questionable (Holling, 1961). In natural systems the shape will be multiform, particularly due to the presence of alternative prey species and different prey stages. In the following subsections the factors affecting the functional response of *T. occidentalis* females to the density of *T. urticae* eggs and males will be considered one by one.

5.2.1 *Adaptation of predators to a given prey density*

If predators are deprived of food for a considerable time and are subsequently transferred to a place with a certain prey density, the prey mortality does not immediately become constant. Figure 22 demonstrates that prey mortality decreases gradually for some time until a level is reached, which is characteristic for each prey density. If the prey individuals are males, the production of webbing prevents constancy. According to Fig. 22 the mean gut content rises to a certain level, dependent on prey density, and in the case of males decreases with the webbing density. The mean hunger of the predators, determining the prey mortality and the predation rate, reaches a new level when the prey density has changed, and this process needs some time. The process of adaption was fully recognized and discussed for fishes by Rashevsky (1959), who distinguished instantaneous rates of consumption from average rates. However, in general no attention has been paid to this phenomenon by experimenters studying predation.

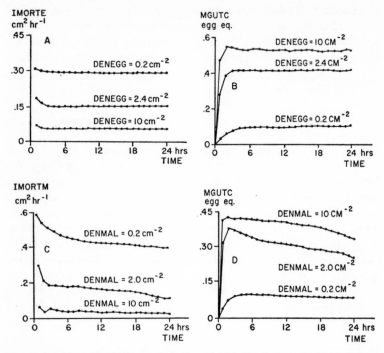

Fig. 22 | Adaptation of starved predators as indicated by the intrinsic rate of prey mortality and the mean gut content during a 24-hour period at different densities of eggs (A, B) and males (C, D). Computed by the program COMPOUND.

Adaptation periods are never mentioned, although they will have a considerable influence on the predation rate during short observation periods. Because the length of the adaptation period increases with decreasing prey densities, adaptation may cause an overestimation of the predation rate at the lower prey densities. This may have contributed to the shape of the functional response curves obtained empirically by Haynes & Sisojevic (1966). They found relatively high predation rates in hungry and fully fed spiders (*Philodromus rufus*) at low densities of fruit flies when observed for three hours, but not for twenty-four hours. Their 'fully fed' spiders might have been hungry as well, since they did not feed for two days before the observations.

92

The influence of adaptation on the functional response to prey density in *T. occidentalis* is shown by the difference between the predation rates during the first and the last six hours of a 24-h period in Figs 16–18. This is, however, partially obscured for males by the difference in webbing density. Nevertheless, the curves for the period 0–6 h show the tendency to have a higher level than the curves for 18–24 h, particularly at the lower prey densities. This is not caused by the difference in webbing density alone, as is demonstrated by the curves for 0–6 h and a high initial webbing density of about 0.36 (Figs 17 and 18A); these curves still have a higher level. The low values of these curves in the range of the lowest prey densities are the result of the value of the initial webbing density, which is exceptionally high for these low prey densities.

5.2.2 *Hunger of the predator*

The gut content exerts an important influence on the predation by *T. occidentalis*, being the main determinant of the success ratio. In Fig. 23 the proportion of successful encounters, as given by the analysis of continuous observations, is plotted as a function of the gut content without regard to other determinants. For eggs and males the relationships can be represented by curves with a decreasing negative slope. For males the success ratio soon approaches zero, but even at the maximum gut content the predator may catch males with a low, but not a zero success ratio. Apparently hunger does not affect the predation rate by a threshold value of the gut content and predetermined digestive pauses as emphasized by Holling (1966), but merely by raising the pro-

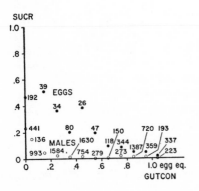

Fig. 23 | The success ratio at different values of the gut content for captures of eggs (solid circles) and males (open circles). The numerals indicate the number of encounters observed.

bability of a capture when a prey is encountered, in the way assumed by Rashevsky (1959).

The influence of hunger may actually include changes in awareness, responsiveness to prey, and success of attack, but the very short and constant distance of reaction of *T. occidentalis* makes a distinction difficult and, as far as the predation rate is concerned, irrelevant. In Holling's praying mantid-fly system, in which awareness is indicated by a movement of the predator's head and responsiveness by a movement of the legs, the success ratio of attacks is always 0.63. In the praying mantid hunger does not affect the success ratio, but only the maximum distance of reaction, and hence the coincidence in space with the prey. A prey is attacked as soon as it comes within this range, but the distance of reaction is zero beyond a threshold value of the gut content. In this way the concept of a digestive pause is prompted by the particular behaviour of the praying mantid, and perhaps also by Holling's low number of replicates (12) of experiments to determine the maximum distance of reaction at each hunger level. Low attack ratios cannot be concluded from the result of twelve encounters. Possibly the digestive pause is not a general phenomenon; anyway the concept cannot be used in the study of predation by *T. occidentalis*.

Sandness & McMurtry (1972) measured the digestive pause as the time between the abandonment of a prey and the capture of the next, comprising Holling's digestive pause, searching time, and pursuit time. They continuously observed *Amblyseius largoensis* females preying on *Oligonychus punicae* females (both mite species) and found a rather cyclic course of successive digestive pauses, series of long pauses folowing series of short ones. They assumed the existence of many hunger thresholds for components of the predatory behaviour of *A. largoensis*, but did not measure the gut content. The time elapse between captures will depend on the gut content in this mite species, as it does in *T. occidentalis*, but the assumption of obligatory digestive pauses operating by thresholds seems to be a too deterministic conception of a predation process among mites. *T. occidentalis* catches prey encountered even at a low hunger level, but with a low success ratio. This is a valuable property of this species as a controlling agent, because it increases its effectiveness at high prey densities.

A positive correlation between hunger and feeding time has been reported for mites (Putman, 1962; Sandness & McMurtry, 1970, 1972) and spiders (Haynes & Sisojevic, 1966). In our system such a relationship is clearly indicated in Tables 10 and 11 and Fig. 14. According to Section 3.3 the ingestion rate is also positively correlated with hunger. The result of both effects is a better prey utilization at a high hunger level. The average amount of food in egg equivalents eaten per prey at different mean gut contents was derived from the predation rates (Figs 16A and 17), the biomass consumed per hour (Fig. 20), and the mean gut content (Fig. 21) at different prey densities. These values have been plotted in Fig. 24. To obtain a value for feeding on males at a high gut content, the ratio of the male utility (the instantaneous value of the amount of food ingested per male per hour) and the intrinsic rate of mortality (the instantaneous value of the proportion of males killed per hour) was derived from Figure 26C for the prey density of sixteen eggs and ten males per cm^2, and also plotted in Fig. 24.

According to Fig. 24 the prey utilization is clearly hunger dependent. The predator simply tries to fill up its gut to the maximum content but, as the line of maximum restoration indicates, it does not succeed. While feeding it is increasingly restrained by the limited availability of food, and its feeding drive vanishes depending e.g. on the palatability of the food (see also Fig. 14). In many cases the prey is not eaten completely, which has also been observed in *T. occidentalis* by Lee & Davis (1968). Figure 24 demonstrates, that at all hunger levels a prey egg provides more food than a prey male.

The searching behaviour and activity may be hunger dependent in predatory insects, as reported for the coccinellid *Adalia decempunctata* by Dixon (1959). In *T. occidentalis* the mean resting time slightly

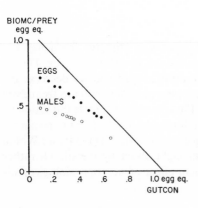

Fig. 24 | The relation between prey utilization and the gut content of the predator for eggs (solid circles) and males (open circles), as derived from results of the program COMPOUND. The line indicates the maximum utilization possible.

increases with the gut content (Tables 8 and 10), which decreases the walking activity. The fact that hunger has little influence on the loco-motory activity of *T. occidentalis* seems to be disadvantageous. Walking is likely to increase the risk of death, and at a high gut content it is a waste of energy. Probably this species has been adapted to a very low prey density, where a high walking activity and responsiveness to prey have 'survival value' even at a low hunger level, because the time elapsed before encountering the next prey will be relatively long on the average. It may be necessary for the predator to maintain a high mean velocity.

The coincidence in space is not affected by hunger, indicating that the predator does not search purposely on areas preferred by the prey. Correlation analysis revealed an effect of hunger on the sensibility for disturbance (Table 8), but a comparison of Tables 11 and 12 shows that it is unimportant.

During predation hunger can fluctuate widely in *T. occidentalis* within short time-periods, since the gut content is halved every 1.59 hours by digestion. In most studies of the effect of hunger on components of predator behaviour, different hunger levels have been created by depri-vation of food over several days, e.g. in studies on mite predation of Mori & Chant (1966) and Sandness & McMurtry (1972). However, in mites the gut will be almost empty after two days under such con-ditions, so the differences in hunger level after several days of food deprivation will be negligible. The differences in behaviour observed under such conditions will not be the result of hunger, but rather of inanition.

5.2.3 *Inhibition by prey*

A decreasing responsiveness with increasing stimulus frequency has been described as a general phenomenon in arthropods by Thorpe (1962). This process, called 'habituation', was found to decrease the attack ratio, when predatory insects were evoked to react to frequently presented artificial prey (Holling, 1966; Wolda, 1961). Habituation or inhibition by prey is thought by Holling to be invoked by the responses of the predator. However, ethologists such as Hinde (1954) and Wolda (1961) argue that habituation is brought about by the stimuli rather than by the responses, being a mere effect of neurophysiological sti-mulus saturation.

Inhibition by prey seems to be commonplace in *T. occidentalis*, since both the success ratios and the disturbance by prey males are negatively correlated with the frequency of contacts with the prey (Table 8).

5.2.4 *The density of the webbing cover*

The webbing produced by spider mites may have several functions:
1 facilitation of locomotion and dispersion
2 regulation of the microclimate
3 indication of the state of depletion of the food resources and the conditioning for social interactions (Wynne-Edwards, 1962)
4 exclusion of competing species (Foot, 1963)
5 protection from predation.
The protection from predation can be threefold: predators may be barred from webbing covered areas, their activity and speed of locomotion may be reduced, and they may be outrun more easily, promoting the escape of the prey.
A negative influence of webbing on predation by phytoseid mites has been reported for adults of *Amblyseius longispinosus* (Mori, 1969), *A. largoensis* (Sandness & McMurtry, 1970, 1972), *Typhlodromus caudiglans* (Putman, 1962), *T. pyri*, *T. caudiglans* and *T. hibisci* (McMurtry et al., 1970), and for larvae of *T. tiliae* (Dosse, 1956), while the adults of the last species seem not to be affected. *T. occidentalis* is said to be attracted by areas covered by webbing, where it would be most effective in destroying phytophagous mites, and where it would be locomotory inactive (Flaherty & Huffaker, 1970b; Huffaker et al., 1969; McMurtry et al., 1970).
Table 8 represents the correlations of behavioural components with the relative density of the webbing cover, as found in *T. occidentalis* in this study. The walking activity is reduced by the webbing, which affects the mean period of walking. This might well be a negative influence due to an increased resistance for locomotion. The coincidence in space is reduced, indicating some barrier effect. The predator velocity is negatively correlated, and the male velocity positively with the webbing density, but there is no distinct influence on the success ratios. The feeding on eggs, which lie under the webbing cover, seems to be extended by webbing. The handling of males is reduced, probably because the webbing interferes with turning and palpating. The males are activated by webbing, and they prefer to walk on it, which suppresses

Table 13 Mean values of behavioural components at different densities of prey males and webbing cover during the period 0–6 hour

	0.2	0.8	5.0	15.0	20.0	0.2	0.8	5.0	15.0	20.0
DENMAL	0.07	0.20	0.36	0.34	0.36	0.04	0.13	0.31	0.30	0.34
GUTCON	0.05	0.13	0.23	0.29	0.38	0.36	0.36	0.38	0.39	0.44
DENWEB										
RESTIM	0.05	0.03	0.03	0.05	0.01	0.04	0.03	0.03	0.06	0.01
ACTIM	0.04	0.05	0.07	0.05	0.07	0.05	0.05	0.08	0.06	0.07
VELPRE	1.10	0.87	0.75	0.76	0.81	0.79	0.80	0.80	0.82	0.85
ACTMAL	0.72	0.72	0.65	0.70	0.58	0.81	0.79	0.71	0.70	0.56
VELMAL	1.00	1.02	0.55	0.30	0.79	1.08	1.00	0.42	0.30	1.00
DISTUR	0.64	0.57	0.39	0.65	0.61	0.63	0.56	0.38	0.64	0.63
COINAA	4.29	3.66	2.85	2.39	1.19	1.57	1.48	1.30	1.02	0.07
COINAR	4.01	3.75	3.57	3.58	3.66	3.63	3.64	3.65	3.67	3.71
COINRA	1.58	1.58	1.58	1.58	1.58	1.58	1.58	1.58	1.58	1.58
SUCRAA	0.033	0.026	0.011	0.007	0.006	0.035	0.029	0.014	0.009	0.006
SUCRAR	0.031	0.021	0.012	0.005	0.004	0.033	0.027	0.016	0.007	0.005
SUCRRA	0.016	0.007	0.002	0.002	0.002	0.019	0.012	0.003	0.003	0.002
MAFTIM	0.15	0.09	0.08	0.07	0.08	0.18	0.12	0.08	0.08	0.07
MAHTIM	0.25	0.15	0.12	0.12	0.11	0.25	0.18	0.11	0.11	0.11

Table 14 Mean values of behavioural components at different densities of prey males and webbing cover in the presence of 16 prey eggs per cm² during the period 0–6 hour

	0.2	0.8	3.2	15.0	20.0	0.2	0.8	3.2	15.0	20.0
DENMAL	0.44	0.39	0.41	0.41	0.46	0.04	0.19	0.45	0.44	0.48
GUTCON										
DENWEB	0.05	0.13	0.22	0.29	0.38	0.36	0.36	0.38	0.39	0.44
RESTIM	0.07	0.05	0.03	0.02	0.01	0.04	0.03	0.04	0.01	0.01
ACTIM	0.04	0.05	0.07	0.05	0.07	0.05	0.05	0.08	0.06	0.07
VELPRE	1.10	0.87	0.76	0.76	0.81	0.79	0.80	0.80	0.82	0.84
ACTMAL	0.72	0.72	0.67	0.70	0.58	0.81	0.79	0.71	0.70	0.56
VELMAL	1.00	1.02	0.75	0.30	0.79	1.08	1.00	0.42	0.30	1.00
DISTUR	0.69	0.60	0.43	0.67	0.64	0.63	0.56	0.39	0.68	0.67
COINAA	4.28	3.66	2.97	2.38	1.18	1.57	1.48	1.31	1.01	0.08
COINAR	4.01	3.75	3.58	3.58	3.66	3.63	3.64	3.65	3.67	3.71
COINRA	1.58	1.58	1.58	1.58	1.58	1.58	1.58	1.58	1.58	1.58
COINAE	0.78	0.49	0.42	0.16	0.37	0.00	0.07	0.49	0.30	0.41
SUCRAA	0.017	0.017	0.012	0.003	0.002	0.035	0.026	0.007	0.002	0.001
SUCRAR	0.006	0.009	0.009	0.000	0.000	0.033	0.023	0.005	0.000	0.000
SUCRRA	0.001	0.002	0.001	0.001	0.001	0.019	0.008	0.001	0.001	0.001
SUCRAE	0.020	0.022	0.021	0.021	0.020	0.040	0.031	0.030	0.020	0.019
MAFTIM	0.08	0.08	0.08	0.08	0.08	0.18	0.10	0.08	0.08	0.09
MAHTIM	0.15	0.13	0.13	0.12	0.12	0.25	0.14	0.12	0.12	0.12
EFTIM	0.05	0.03	0.02	0.32	0.23	0.10	0.07	0.06	0.30	0.25
EHTIM	0.07	0.05	0.03	1.39	0.50	0.12	0.08	0.07	1.10	0.62

their feeding on the leaf's epidermal cells.

The actual influence of the webbing on predation is demonstrated by Figs 17 and 18, and Tables 13 and 14, showing the influence of an initial relative webbing density of about 0.36 during the first six hours after the release of the predator. In general the predation rate is reduced at this webbing density, especially for eggs at low male densities. The influences on the coincidence in space and the predator's velocity are most distinct, whereas the influences on the other components can hardly be traced. The conclusion is that webbing has a distinct barrier effect, especially with regard to active prey and eggs. It hinders *T. occidentalis* sufficiently to effect the predation rate. The higher coincidence in space with eggs at high male densities may be because of the restriction of the predator to areas that are relatively devoid of webbing.

Figure 25 depicts the webbing density at the end of a 24-h period as a function of the male density. It is shown that the webbing density levels off at about 3 males/cm^2; it then rises again with a low slope due to a relative constancy or even increase of the male velocity, which declines sharply at densities below 5 males/cm^2 (Table 11). In restricted areas the males produce a rather uniform webbing cover. This phenomenon may be related to intrinsic optimum conditions for spider mite colonies, but it surely provides protection against predation by *T. occidentalis*. If this species is common in spider mite colonies, it is more so by the effect of trapping and a high reproduction rate in the presence of much food, than by attraction.

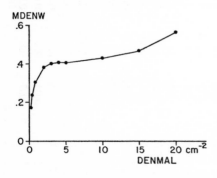

Fig. 25 | The mean relative density of the webbing cover in the period 18–24 hour as related to male density. Computed by the program COMPOUND.

5.2.5 *Subsidiary effects of the prey male density*

The frequency of mutual, bodily contacts or olfactory stimuli emanated by the spider mite males affect several components of their behaviour. This interaction between males leads to clustering and a decrease of their locomotory activity and velocity with increasing density, thus reducing the rate of encounters between the predator and males.

Incoordinate aggregations are quite common in many animal species (Gervet, 1968). Webbing aggregations and sexual behaviour may contribute to the phenomenon in spider mite males. Aggregation was also observed, however, in females by Mori (1969). He considered this to be, at least partially, the cause of a decline in the predation rate of *Amblyseius longispinosis* at high prey densities. Fleschner (1950) pointed out that prey aggregation can enhance predation, if the predator reacts to encounters by a more intensive search of the area. This behaviour was never observed in the mite system.

A factor increasing the predation rate is the disturbance by males. A high percentage of contacts between a resting predator and active males results in reactivation of the former (Table 11), thus raising the encountering rate. The mean walking period is also extended by encounters, the predator being easily frightened and upset by the males. Simulation reveals this effect to have a distinct influence on the predator's activity (compare the values of the mean resting and walking periods in Tables 10 and 11).

Disturbance of feeding predators may increase the success ratio if the prey, bumping into the predator, is immediately attacked (Haynes & Sisojevic, 1966; Sandness & McMurtry, 1970). *T. occidentalis*, however, is not easily disturbed while feeding, like *Amblyseius longispinosus* (Mori, 1969), and it does not return to abandoned prey as reported for *Amblyseius largoensis* (Sandness & McMurtry, 1972) and *Typhlodromus fallacis* (Ballard, 1954). Such a behaviour would easily be interrupted by active prey at high densities. Probably, the responsiveness of feeding predators to disturbance depends on the size of the disturbing prey.

5.2.6 *Consequences of polyphagy*

Predators are rarely monophagous and even if they are, the different developmental stages of the prey represent different kinds of prey.

Thus, for *T. occidentalis* the eggs of *T. urticae* are more attractive and provide more food than the males. The latter, however, are in general more frequently encountered. To indicate the prey characteristics of different kinds of prey de Ruiter (1956) introduced the term 'prey-value', which he defined as the frequency of feeding responses. This is a function of the attractiveness of the prey and the rate of discovery. As demonstrated by Fig. 18 the presence of alternative kinds of prey can influence the shape of the functional response to the density of one kind of prey. Hence the 'prey-value' of one kind of prey can be affected by the density of other kinds of prey. This is especially so, if such 'prey-values' depend on the hunger level of the predator. This section discusses the influence of the gut content of *T. occidentalis* on the 'prey-values' of *T. urticae* eggs and males, and the implications of such relationships for the stability of prey density.

The term 'prey-value' is rather ambiguous, because the prey has different values when considered from the viewpoint of the prey population or the predator population. Therefore we will use two new measures of 'prey-value'.

From the viewpoint of the prey population predation is a cause of mortality. When N_1 denotes the prey density, N_2 the predator density, and t time, the rate of mortality caused by predation can be expressed by

$$dN_1/dt = - \mathbf{IMORT} \times N_1 \times N_2$$

where $\mathbf{IMORT}$ represents the intrinsic rate of mortality of the prey population caused by predation. $\mathbf{IMORT}$ has the dimension $h^{-1}/(predator/cm^2)$, which reduces to $cm^2 h^{-1}$ when it is conceived as the rate of increase of the area cleared of prey by one predator (compare the 'area of discovery' used by Nicholson (1933), which is the ratio of the number of eggs laid by an insect parasite and the density of its host (Tinbergen & Klomp, 1960)).

For the mite system $\mathbf{IMORT}$ is computed by the program $\mathbf{COMPOUND}$ for different prey densities. Every $\mathbf{DELT}$ the number of prey killed in $\mathbf{DELT}$ is divided by the actual number of prey on the leaf disk, and the predator density (0.2). The mean steady state values (in the period 18–24 h) are plotted in Fig. 26. According to this figure $\mathbf{IMORT}$ decreases with increasing prey density. In Section 5.1 it was concluded that prey density influences the conditions created by prey behaviour (aggregation and webbing density) and the hunger level of the predators (the gut content), which all determine the predation rate

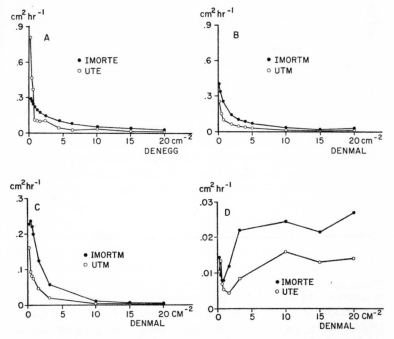

Fig. 26 | The intrinsic rate of prey mortality and prey utility during the period 18–24 hour as related to prey density; for predation on eggs (A), males (B), males when $DENEGG = 16/cm^2$ (C) and eggs when $DENEGG = 16/cm^2$ and the male density is varied (D). Computed by the program COMPOUND.

and hence IMORT. In both ways a change of the density of one prey can influence the IMORT of alternative kinds of prey. Obviously this will be most common when the IMORT of all kinds of prey depends on the hunger level of the predators. Because this has interesting consequences in case of polyphagy, the dependency of IMORT on the gut content of the predators in the mite system will be considered more closely.

The steady state values of IMORT at different densities of eggs and males can be plotted against their concomitant values of the gut content given in Tables 10 and 11, as computed by COMPOUND (Fig. 27). Within the range of the prey densities used the values almost lie on

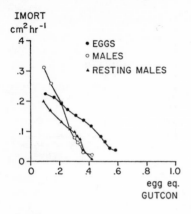

Fig. 27 | The relationship between the intrinsic rate of prey mortality due to predation and the gut content of the predators, as derived from the results of the program **COMPOUND**.

straight lines with a negative slope, indicating a negative correlation. We see that at a low gut content the I MORT of males is higher than the I MORT of eggs, but at a higher gut content the situation is reversed. Because the gut content is correlated with prey density, the relationship for males in Fig. 27 is biassed by the influence of aggregation and webbing density. To compare the unbiassed influences of GUTCON on I MORT of eggs and males we may consider resting males only, to which applies at the same value of GUTCON:

I MORT of resting males

$$= \frac{COINAR \times SUCRAR}{COINAE \times SUCRAE} \times IMORT \text{ of eggs,}$$

with the value of COINAR on a leaf disk without webbing and aggregations. COINAE, COINAR and SUCRAR are given by Tables 10 and 11, IMORTE and SUCRAE were obtained for the same values of GUTCON by linear interpolation in Fig. 27 and Table 10. Using the COINAR value at the lowest male density (COINAR does not change very much anyway), the IMORT of resting males could be derived and plotted in Fig. 27. The lines through the points for normal males and resting males demonstrate the importance of the prey mortality reduction by webbing and aggregation. It is clear that these influences have to be excluded when the influence of the gut content is considered.

The lines through the points for eggs and resting males in Fig. 27

demonstrate the pure influence of predator satiation on the prey mortality. They run more or less parallel, but it is clear that the IMORT ratio of the two kinds of prey, the 'relative prey-value' of de Ruiter (1956), changes rapidly with the hunger level. Such a dependency of the relative prey-value on the state of nutrition has also been observed in salticid spiders, as mentioned by de Ruiter.

Murdoch (1969) introduced the term 'switching' for the phenomenon, that a change in the ratio of the densities NA and NB of two alternative prey species A and B is accompanied by a change in the ratio (NPA/NA)/(NPB/NB), where NP denotes the number killed per hour. When NP is considered as an instantaneous value, as it should be, the last ratio equals IMORTA/IMORTB. According to this definition of 'switching', the decreasing IMORTM/IMORTE ratio with increasing gut content, demonstrated by Fig. 27, makes it plausible that in a mixed prey population *T. occidentalis* will 'switch' to its preferred prey species, when the prey density increases. This is plausible, because at any composition of the prey population the mean gut content will increase with the prey density, and a change of the prey density in most cases will imply a change in the ratio of the densities of the prey species (although this is not necessary for hunger dependent switching). For the discussion of the consequences of this phenomenon it will be helpful to introduce a second measure of 'prey-value'.

From the viewpoint of the predator population the prey population is a source of food, which is ingested per cm^2 at a certain rate:

$$dI/dt = UT \times N_1 \times N_2$$

where N_1 now represents prey density expressed in prey equivalents per cm^2, N_2 predator density, and UT the intrinsic rate of ingestion, to be termed the prey utility. UT has the dimension $h^{-1}/(predator/cm^2)$, which reduces to cm^2h^{-1} when it is conceived as the rate of increase of the area cleared of prey biomass by one predator. UT and IMORT are not equivalent, because the prey killed is not always utilized completely. The steady state values of UT are also plotted in Fig. 26. In the figure, UT sometimes outnumbers IMORT. This is because the UT values used are computed by COMPOUND just before prey replenishment, those of IMORT just after.

In the steady state of the mite system, when the mean gut content is constant, the ingestion rate equals the digestion rate. Let GUTCON denote the constant mean gut content of the predators and DIGEST the digestion rate per prey equivalent (see Section 3.4), then the ingestion rate equals $GUTCON \times DIGEST \times N_2 = UT \times N_1 \times N_2$. This was already appreciated by Rashevsky (1959). We have:

$UT = GUTCON \times DIGEST/N_1$.

For eggs we have:

$UTE = GUTCON \times DIGEST/BIOEGG$

and for males:

$UTM = GUTCON \times DIGEST/BIOMAL$,

where BIOEGG and BIOMAL denote the prey densities in biomass per cm^2. DIGEST is assumed to be equal for eggs and males. Therefore, at the same value of GUTCON the relative utility of the males can be represented by:

$$RUTM = UTM/UTE = BIOEGG/BIOMAL = DENEGG/DENMAL$$

To demonstrate the relationship of RUTM and GUTCON without bias from the differences in webbing density and aggregation, we will consider resting males only. The density of resting males is equivalent to the density of normal males (i.e. will result in an equal gut content of the predators), when the two categories of males have the same mortality rate. This rate is proportional to the density of the males at a certain value of GUTCON. Hence, to obtain the density of resting males we have to multiply DENMAL by the ratio of IMORT for normal males and resting males, which can be derived from Fig. 27. RUTM for resting males was calculated for the GUTCON values computed by COMPOUND for the different densities of males, both given in Table 11. The corresponding values of DENEGG were obtained by linear interpolation in Table 10. RUTM for resting males is plotted in Fig. 28 as a

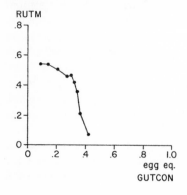

Fig. 28 | The relationship between the relative male utility (relative to the egg utility) and the gut content of the predators, as derived from the results of the program COMPOUND.

function of GUTCON, which reveals a clear-cut hunger dependency. If the hunger decreases, rather abruptly the males become less important as a food source.

T. occidentalis apparently tends to switch its main prey species according to its hunger level. This phenomenon is related to a hunger dependent difference in prey attractiveness (see the relationship between the success ratios and the gut content in Fig. 23) and to a different maximum of the mean gut content reached for each single species of prey. Differences in preference are often reported in studies on mite predation of alternative prey species or stages (Burnett, 1970; Chant, 1963; Collyer, 1958; Dosse, 1956; Elbadry et al., 1968; Flaherty & Huffaker, 1970), so switching may be common in mites. This emphasizes Chant's remark (Chant, 1961a) that in studies of mite populations the determination of predator/prey ratios has little value, if nothing is known about the composition of the prey population.

Switching in invertebrate polyphagous predators may be common. Murdoch (1969) studied switching by marine drilling whelks, preying on mussels and barnacles. Previously trained whelks switched to the most abundant species where no strong preference existed, a factor contributing to the stability of the population system. In a later paper Murdoch (1973) again stresses the possibility of stabilization by switching to alternative prey, but he does not mention the hunger factor. In his example he uses Holling's disk equation for the description of the functional response curves. Such curves level off because of a reduced searching time/handling time ratio at high prey density. In nature, a limitation of the predation rate by satiation may be at least as common as a limitation by time, in particular when the handling time per prey is short. Therefore hunger dependent switching may be of general interest, because it can contribute to the stability of the density of the prey species preferred. The effective density of two alternative prey species A and B can be expressed as $NA + RUTB \times NB$. When NA increases, the gut content of the predators increases but $RUTB$ may decrease. In that case the effective prey density and the gut content increase less than they would have done otherwise, and the mortality rates remain higher. At a high density of alternative prey the mortality rate of a preferred prey species will be buffered.

Hunger dependent switching by itself cannot induce regulation, because regulation requires an increase of the prey mortality rate at an increasing prey density. $RUTB$, however, decreases only when the gut content of the predators increases, and hence the mortality rates decrease. Nevertheless it is an important phenomenon. When the prey species preferred becomes abundant, the predator tends to delimit itself to this species. Such predators can be excellent controlling agents, since they combine the quality to kill a high number of the prey preferred at a high density of this species, with the ability to survive on alternative prey species when the prey preferred is scarce. It is, however, absolutely necessary that the species to be controlled has the highest preference at a low hunger level. If not so a high prey density causes the predator to neglect the species in 'favour' of others.

Switching in invertebrate polyphagous predators has received little attention in biological control. It can be advantageous to select a controlling agent with respect to the preference of all alternative prey species at different hunger levels. The most efficient predator is often not the most important one in reducing high prey numbers (e.g. Fleschner, 1950), which sometimes may be an effect of switching. Very effective will be a predator, which switches to adult females at a low hunger level. It seems likely, that the stability of arthropod populations in complex biocoenoses as mentioned by Voûte (1946a, b) is not only effectuated by the combined effect of parasites and predators (Tinbergen & Klomp, 1960), but also by the presence of alternative prey species and hunger dependent prey preferences.

5.2.7 *The interval of prey replenishment*

In most models for predation the prey density is represented by a constant. As argued by Royama (1971) this is an unrealistic assumption, since captured prey individuals are not replenished immediately. In fact the actual prey density is a variable. It is treated as such in the simulation models presented in this book.

In experimental studies of the functional response to prey density, prey replenishment is often omitted, e.g. Bravenboer (1959), Haynes & Sisojevic (1966), Mori (1969) and Mori & Chant (1966). Chant (1961b) and Sandness & McMurtry (1970) replenished dead prey daily. Only continuous observation of an experimental system enables immediate replenishment (Beukema, 1968).

To evaluate the significance of prey replenishment, the mean actual prey density as given by COMPOUND was computed as a proportion of the prey density maintained. These values are plotted in Fig. 29

108

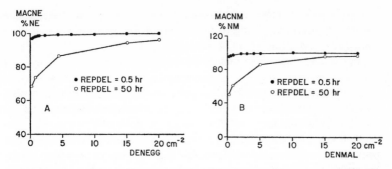

Fig. 29 | The influence of prey replenishment on the mean actual prey density in the period 0–6 hour as a percentage of the prey density maintained. Computed by the program COMPOUND for predation on eggs (A) and males (B).

against the prey density maintained for a thirty-minute replenishment interval and no replenishment at all. The thirty-minute interval used by Kuchlein (in prep.) gives for all densities sufficiently small deviations of the actual density, whereas without replenishment the predation rates would have been underestimated at low prey densities. Figures 16B and 17 show that this underestimation would have been negligible. But in general in predation studies of responses to prey density without prey replenishment it is necessary to consider the actual prey density in the lower range.

5.3 The role of chance in the predation process

In Sections 4.2 and 4.3 the theoretical differences between deterministic, stochastic, and compound simulation models are expounded. It has been pointed out that deterministic models for simulation of stochastic processes can be erroneous. Since different models are applied to simulate predation on *T. urticae* eggs by *T. occidentalis* females, it is possible to determine, whether the error due to deterministic simulation can be seen in the results of simulation of a more tangible predation process. Have the theoretical differences any practical significance? The deterministic model gives for all prey densities a considerably lower predation rate (Fig. 16A), a higher consumption of biomass per hour (Fig. 20A), and a higher mean gut content (Fig. 21A) than the stochastic model. The results of the program for compound simulation are intermediate, but close to the results of the stochastic model. Two sources of deviations are conceivable:

a In the deterministic model the success ratio at the mean gut content will be somewhat lower than the mean success ratio in a stochastic process, firstly because the mean gut content is higher in the former, and secondly because the relationship between this ratio and the gut content is concave (see Fig. 23).

b The rate of abandonment will be lower in the deterministic model, because the relationship between this rate and the gut content is concave (see Fig. 14A). This implies that the proportion of predators handling prey, and hence ingestion, is higher.

In some respects the deviations may be additive. For instance, the deterministic model suggests that the prey killed is always ingested completely, since the number killed per hour almost equals the biomass consumed per hour (compare Figs 16A and 20A). The other models indicate an average consumption of about one third of the prey killed at the higher prey densities.

Curvilinear relationships of the success ratio and the rate of abandonment with the gut content may be common in many predator species. Therefore, deterministic models for the simulation of predation processes have to be used tentatively. It seems indispensable in the study of predation to take account of the role of chance.

5.4 Implications of the functional response derived

5.4.1 *The numerical response to prey density*

Laing & Huffaker (1969) brought a population of *T. urticae* on strawberry plants into a greenhouse and introduced *T. occidentalis* to this population. They observed increases in the density of the predator immediately following peaks in the prey density. This numerical response to the high prey density may be caused by a reduction of the predator mortality, by a higher reproduction rate, or by both.

Mortality due to starvation requires about three days of food deprivation. Consequently the prey density has to be extremely low to effect such a mortality. It seems more reasonable to assume that the numerical response of *T. occidentalis* is the result of a variable reproduction rate. This is confirmed by the numbers of eggs laid per 24 hour by ovipositing predators at different densities of prey eggs (Table 6) and by the fact that they produced less eggs the day next to being transferred to empty leaf disks. An increasing oviposition rate of *T. occidentalis* with rising prey density was also observed by Chant (1961b). The same has been reported for other phytoseiids by Bravenboer & Dosse (1962), Burnett (1970), Herbert (1956) and Smith & Newson (1970), and for an aceosejid mite by Rivard (1962).

Because the absorbtion rate of food can be assumed to be proportional to the gut content, it seems also reasonable to assume that the oviposition rate in predatory mites is approximately proportional to the gut content. When the numbers of eggs laid per 24 hour, given in Table 6, are plotted against their concomitant values of the mean gut content of ovipositing predators given in Table 7, then they almost lie on a straight line with the formula:

oviposition rate = $4.38 \times$ gut content $- 0.66$ (Fig. 30).

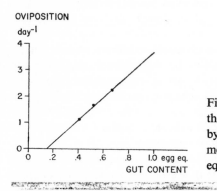

Fig. 30 | The relationship between the number of eggs laid per 24 hour by *T. occidentalis* females and their mean gut content in *T. urticae* egg equivalents.

According to Fig. 21 the gut content of the predators levels off at a prey density of about five per cm². Therefore the numerical response of *T. occidentalis* to the prey density will be most significant in the lower range of prey density.

5.4.2 *Regulation of spider mite populations*

The number of prey killed and the number of eggs laid per predator per time unit increases in *T. occidentalis* with the prey density. It remains to be seen, whether these responses can increase the mortality rate of the prey population to such an extent that it exceeds the reproduction rate. This requires that the product of the intrinsic rate of mortality $IMORT$ and the predator density increases with the prey density. From Figs 26A–C it is clear that the functional response alone is unable to increase $IMORT$; only the $IMORT$ of eggs is raised at intermediate densities of males (Fig. 26D). A growth of the prey population, however, increases the oviposition rate of the predators. After about one week this would result in an increased density of adult predators and a consequently higher prey mortality rate.

The aim of this study is the analysis of the functional response of *T. occidentalis*, hence the information available for a quantitative description of the numerical response is rather scanty. It is unlikely that the relationship oviposition rate $= 4.38 \times$ gut content $- 0.66$ gives a complete description of the oviposition rate at different prey densities. This relationship may be curvilinear at low and high values of the gut content. Moreover, the density of the predators may affect their oviposition rate (Kuchlein, 1966). Nevertheless, the relationship will be used in combination with the functional response to describe a rough total response to prey density.

The total response can be defined as the relationship between the mortality rate of the prey caused by the next generation of predators and the current prey density. The mortality rate of the prey, to be termed prey risk, equals $IMORT$ times predator density. Starting from a predator density one, the predator density of the next generation will be equal to the ratio of the actual oviposition rate and the oviposition rate that keeps the predator density constant.

The predator density is constant, when each adult female produces on the average one adult female in the next generation. This implies that only two of her eggs laid will develop to adults, when the sex ratio is one. The total number of eggs laid by a female is o times l, where o is the number of eggs laid per day and l the longivity of adult females in days. Hence, the proportion of animals surviving the preadult stages (s) must be $2/(ol)$. The predator density will remain constant when the oviposition rate o equals $2/(sl)$ eggs per predator per day. The density of the next generation of predators will be $o/(2/(sl)) = sl/2 \times o = q \times o$, where q is the expectation value of the reproductive period. When we exclude any predator mortality factors other than senescence, s is one and l is about twenty-five days (Kuchlein, 1966). Then q equals 12.5 days. For the different densities of eggs and males with their concomitant

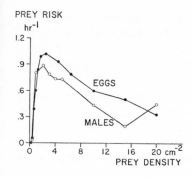

PREY RISK
hr^{-1}

Fig. 31 | The total response of *T. occidentalis* to the density of *T. urticae* eggs and males, as indicated by the instantaneous rate of prey mortality when the next predator generation is adult, the current density of the predator population is one, and senescence is the only cause of predator mortality.

PREY DENSITY

values of the gut content given in Tables 10 and 11 the oviposition rate can be computed. The values of I MORT are also known (Figs 26A and 26B). Hence the prey risk caused by the next generation of predators can be computed as I MORT × oviposition rate × 12.5. The total response to the density of eggs and males derived in this way is depicted in Fig. 31.

The total response curves show a steep rise to a maximum at about two prey per cm^2. In spite of the restrictions about the reliability of the numerical response incorporated, the curves indicate a strong possibility of regulation at a low prey density within one generation period. Because at the higher prey densities the predator population will continue growing, even extremely high prey densities will finally be brought back by a succession of predator generations. This process, however, is rather delayed and cannot prevent damage by noxious prey. However, the 'short term' regulation at a low prey density will be most important for the praxis of biological control.

Whether the total response does imply regulation depends on the reproduction rate of the prey population, the predator density, and the predator mortality. The maximum fecundity of *T. urticae* females is about 38 eggs, which hatch after 6.7 days (Laing, 1969b). The prey population can maintain itself if the number of eggs killed in 6.7 days does not exceed 38-2 per female, so the egg risk (the instantaneous number of eggs killed per hour) must not exceed

$$(38-2)/38 \times 1/(6.7 \times 24) = 0.006 \text{ h}^{-1}.$$

At a predator density one the maximum egg risk exceeds 1.0 h^{-1}. The maximum egg risk is proportional to the predator density, hence even at a predator density as low as 0.006 the mortality rate of *T. urticae* due to predation of eggs can exceed its maximum reproduction rate, which satisfies a necessary condition for regulation. Actually, the minimum predator density necessary for maintaining prey numbers will be lower, because the predation of other prey stages counts also.

As mentioned above the prey risk caused by the next generation of predators equals IMORT times the density of this generation. This density equals the oviposition rate times the reproductive period q. The product $o \times q$ must exceed one at the prey density with the maximum prey risk caused by the next generation of predators, otherwise the predator population cannot exist at prey densities below this value, and regulation will not occur. The maximum prey risk occurs at about two eggs or males per cm^2 (Fig. 31). According to Tables 10 and 11 the mean gut content of the predators at this density is 0.39 and 0.27 egg equiv. respectively, corresponding with oviposition rates of 1.05 and 0.52 per 24 hour (Fig. 30). Regulation requires that $q = sl/2$ will not be lower than $1/1.05 = 0.95$ days for predation on eggs, and $1/0.52 = 1.92$ days for predation on males. Even when the longivity of the adult females is constantly 25 days (which is not likely to be true in natural systems), the percentage mortality in the subadult stages $((1 - s) \times 100\%)$ should not exceed 92% for predation on eggs and 85% for predation on males. Since the potential fecundity of *T. occidentalis* females lies between 50 and 100 eggs, it is clear that a high mortality of preadult stages must already be common due to natural factors. Very easily a subsidiary mortality factor, such as the application of pesticides, can inhibit regulation. It is suggested that a high predator mortality, rather than a low predator density, will impede the occurrence of regulation. If this is true for natural systems, it is not amazing that predacious mites are not able to regulate phytophagous mites in sprayed orchards (McMurtry et al., 1970).

Experiments with populations on detached bean leaves (Chant, 1963) and on strawberry plants (Laing & Huffaker, 1969) confirm the conclusion that *T. occidentalis* is able to counteract increases of the density of *T. urticae*. However, in these experiments the prey density was often reduced below the level where the predator population can maintain itself, when high numbers of predators followed peaks in the prey density. Stable population systems may persist in the presence of alternative prey species only. Then, the effect of switching (see Section 5.2.6) may prevent the dying out of the predator population.

114

5.5   Application of the results in population models

In simulation studies of the interactions of predator and prey populations it may be necessary to compute the predation rate for each of the prey stages present in consecutive time-steps. The model COMPOUND can easily be extended to more than two prey stages or prey species, but as a submodel for population models it has two disadvantages: its complexity and its short time-step for integration. It would be convenient if its essence could be laid down into a concise descriptive model, comprising only a few structural statements without time integrals. This is possible, because the system reaches an equilibrium within a few hours (Fig. 22). In this equilibrium the state variables can be assumed to be constant. The equilibrium is the result of the feedback loops in the predator-prey system (see the diagram in Fig. 10). All predator-prey systems will show such feedback loops. Either the one involving the predation rate and the encountering rate, or the one involving the gut content, success ratio and the predation rate, will induce an equilibrium. For population models with a time increment larger than ten hours, the equilibrium values of the predation rates may be used.

One way to obtain a descriptive submodel is by the application of some kind of multiple regression on data points, which are computed with the complex model. It would be possible to run the program COMPOUND for orthogonal series of prey densities and other decision variables, and to apply polyfactor analysis with the resulting equilibrium predation rates as dependent variables and the decision variables as determinants.

Another method considers the prey densities to be equivalent to a certain density of one prey species, according to the relative utilities of the prey species. We denote the gut content in the steady state by GUTCON, the densities of the prey species present e.g. by DENEGG and DENMAL, and the relative prey utility of the second species by RUTM. RUTM is a function of GUTCON as visualized in Fig. 28. The total density expressed in prey equivalents (TDEN) is always:

$$TDEN = DENEGG + RUTM \times DENMAL$$

For the predator TDEN is equivalent to an equal density of the first prey. The relationship between GUTCON and TDEN is given by the relationship between GUTCON and DENEGG, depicted by Fig. 21A.

The intrinsic prey mortalities IMORTE and IMORTM are related to GUTCON as shown in Fig. 27. The numbers of prey captured per time unit per predator, NPE and NPM, can be computed after solving GUTCON in an implicit loop:

At first GUTCON is set equal to the initial gut content. Then we have:

```
RUTM = f(GUTCON)
TDEN = DENEGG + RUTM × DENMAL
GUTCON = f(TDEN)
```

This sequence of computations is repeated until GUTCON is constant. Then:

```
IMORTE = f(GUTCON)
IMORTM = f(GUTCON)
NPE = IMORTE × DENEGG
NPM = IMORTM × DENMAL
```

The functions f may be given as pairs of coordinates for interpolation. Other prey species can be included by expressing their prey values all relative to the first species.

In the last method all necessary information is obtained from the relationships of the gut content and the predation rate with the densities of single species only (see Section 5.3.6). Therefore, the influence of the density of one kind of prey on the predation of another, for example by disturbance of the predator, is not incorporated in the functions used. Where such interactions are not important, the method may be a useful tool to estimate the numbers of each species killed in a mixture. The essential relationships have to be obtained by simulation or experiments for single species only, which saves much labour and money. In experiments the gut content may be difficult to measure, but in the method with the implicit loop it can be replaced by dependent measures like the oviposition rate or the success ratio.

Summary

Predacious mites are considered to be important natural enemies of phytophagous mites. Their efficiency in the natural control of prey populations depends on the relationships of the number of prey killed per predator per time unit and the oviposition rate on the one hand and prey density on the other hand. These relationships determine the functional and the numerical response of the predator population to the prey density. The shape of these responses indicates to what extent the mortality of the prey population will be raised by an increase of the prey density. When the mortality rate is increased to such an extent that it exceeds the reproduction rate of the prey population at a certain value of the prey density, a necessary condition for the regulation of the prey would be satisfied. Through the operation of such a regulating mechanism the predator and the prey population may remain present at low levels, which gives protection against outbreaks of harmful prey species.

Curves of the functional response of *Typhlodromus occidentalis* females to the density of *Tetranychus urticae* eggs and males on disks of bean leaf have been obtained experimentally by Kuchlein (in preparation). The curves do not correspond with the fundamental types distinguished by Holling (1959a, 1961). This study is an attempt to explain these curves and to find out on which conditions *T. occidentalis* can regulate prey species such as *T. urticae*. For that purpose simulation models were constructed to compute the relevant components of the behaviour of the predator and the prey, the state variables of the predator-prey system, and the resulting predation rates at different prey densities. In this system approach different system elements are distinguished. The components of behaviour considered are the proportion of encounters between the predator and the prey resulting in a capture (the success ratio), the length of handling periods, the coincidence in space, and the locomotion activity and velocity of the predator and the prey. The state variables considered are the engagement of the predator in searching and handling prey, the actual prey densities, the hunger of the predator, and the density of the webbing cover produced by the *T. ur-*

117

ticae males. A quantitative measure of the degree of satiation of the predator was found in the gut content; a measure for the relative density of the webbing cover in the proportion of grains free from the leaf's surface when the leaf disk is sprayed with a fine powder.

To compute the gut content in the simulation of the predation process it is necessary to know the rate of ingestion when the predator feeds on a prey, and the rate of evacuation of food from the gut (denoted by digestion). Feeding experiments with radioactive prey revealed that the ingestion rate is not constant, but depends on the gut content of the predator and on the availability of food in the prey. The rate is higher when the prey is a *T. urticae* egg, than when it is a male. The maximum gut content was computed to be 1.08 prey egg equivalents, the maximum food contents of eggs and males found are 0.94 and 0.67 egg equivalents respectively. The digestion rate is assumed to be proportional with the gut content. The coefficient of this proportionality, obtained from the restoration of the behaviour during a long digestive pause, measured 0.435 hour^{-1}.

The relationship between the relative density of the webbing cover and the total distance walked by the males on the leaf disk has been defined quantitatively. The production of webbing is gradually reduced, or the males tend to walk along the same pathways.

Leaf disks were prepared with a standardized adult female *T. occidentalis* (with an empty gut and two days old), or with a normal ovipositing female, and a different number of *T. urticae* eggs or males. These disks were observed continuously and the events occurring were recorded chronologically. The locomotion velocity of the predator and the males, and the locomotion activity of the males was measured during this observation. From the records of events a computer program derived for a series of gut-content classes, the mean values of the resting period and the walking period of the predator, the coincidence in space and the success ratio for different encountering situations, the proportion of disturbance in encounters of active males and a resting predator, and the handling and feeding periods.

The velocity of the standardized predators did not differ significantly before and after feeding on the first prey (1.15 and 1.12 m/hour respectively).

Oviposition by the predator slightly increases the predation rate by raising the success ratio, but reduces the locomotion velocity. The success ratio of standard predators was four times higher than the suc-

118

cess ratio in Kuchlein's experiments, which was accounted for in the simulation models.

The multiple relationships between the components of behaviour and the state variables were evaluated by partial correlation analysis and polyfactor analysis. The latter method provided quantitative descriptions of the relevant, curvilinear relationships by an iterative procedure for multiple regression. These descriptions were used in the models for simulation of the predation process.

CSMP programs for three kinds of models are described to simulate a predation process on a computer, each dealing with chance variables in a different way. A deterministic model gives erroneous results as soon as the predation rate is a curvilinear function of stochastic variables, whereas a stochastic model consumes too much computer time. An intermediate approach applies deterministic simulation to classes of the stochastic variables in a hypothetical population of predators. The classes in this method of compound simulation are chosen in such a way that within the classes the relationship with the predation rate is approximately linear. The classes contribute to the expectation value of the output variables on the base of the relative frequency distribution of the predators over the classes. The three types of models were applied to simulate the predation process with prey eggs only. The most comprehensive model for the predation of prey eggs and males was built according to the principles of compound simulation. It computes expectation values of the numbers of prey eggs and males destroyed, the biomass consumed, the gut content and occupation of the predator, the actual prey densities, the instantaneous prey mortality and prey utility, and the relative density of the webbing cover as a function of time, the prey density maintained, and the prey replenishment interval. Compound simulation can be applied to study all sorts of stochastic processes.

The results of simulation lead to the following conclusions with respect to *T. occidentalis* and *T. urticae*:

All functional response curves obtained by simulation deviate from the fundamental types of Holling. These deviations are discussed and explained with the help of the values of the components of behaviour at the different prey densities.

Hunger exerts an important influence on predation. The success ratio decreases with increasing gut content, which relationship is the main determinant of the functional response to prey density. There is no distinct threshold-value of the gut content for the evocation of attacks. Hunger induces a long feeding time and a high ingestion rate, and hence a better utilization of the prey captured. At all hunger levels a prey egg provides more food than a prey male.

In mites, which are deprived of food for a different number of days, the difference in hunger level will be negligible. Differences in behaviour demonstrated by such mites will not be effected by hunger, but rather by starvation.

The success ratio seems to be reduced by a high frequency of encounters with prey. Probably predation by mites can be inhibited at a high prey density by stimulus satiation.

Webbing produced by the prey reduces the encountering rate of the predator and the prey, the locomotion velocity and activity of the predator, and the handling time for prey males. It increases the velocity of the prey males, but in general reduces the predation rate, because it has a distinct barrier effect. Other factors related to prey density are the disturbance by active males and the aggregation of the males. Disturbance reduces the mean period of resting and extends the mean period of walking, increasing the predator activity and the encountering rate. Disturbance of feeding predators does not affect the success ratio. Aggregations of males reduce the rate of encounters between the predator and active males, and the activity of the males. Satiation, prey aggregation and the production of webbing are the main factors determining the shape of the functional response of $T.$ $occidentalis$ to the density of $T.$ $urticae$ males on leaf disks.

The prey mortality rate and the prey utility depend on the gut content of the predator. The ratio of the mortality rate or the utility of different prey species is not constant, but changes with the gut content. This induces switching from one kind of prey to another at increasing prey density.

The influence of prey replenishment on the functional response curves is almost negligible for an exposition time of six hours. In general the predation rate will be underestimated for low prey densities when the prey captured is not replenished in experiments.

The number of eggs laid per 24 hour by the predator is about four times its mean gut content. This linear relationship between the oviposition rate and the gut content implies a numerical response of the predator population to prey density, which is most intensive at prey densities below five eggs or males per cm^2.

120

The combined effect of the functional and the numerical response to prey density is indicated by the prey risk induced by the next generation of predators as a function of current prey densities. Such total response curves show a steep rise to a maximum of 2.4 prey eggs or 2.0 prey males per cm^2, and a gradual decline beyond these prey densities. It has been computed that an equilibrium density of at least 0.006 predators per cm^2 at a prey density below 2.4 eggs per cm^2 is a necessary condition for 'short term' regulation of the prey population by predation of eggs. A high predator mortality can inhibit such a regulation. The expectation value of the reproductive period of the newborn predators must not be lower than 0.95 days for predation on eggs, and 1.92 days for predation on males.

Some conclusions derived in this book concern general aspects of predation:
It seems to be indispensable in the study of predation to take account of the stochastic character of predation processes with a single predator. Deterministic simulation models can give incorrect results.
In natural systems the functional responses of predators to prey density will be multiform, and very probably most of them will differ from the fundamental types described by Holling (1959a, 1961).
Predators need some time to adapt their mean gut content to a new prey density. The adaptation time is longer at a low prey density, than at a high one. This may lead to an overestimation of the predation rate at low prey densities, when starved predators are used in experiments with a short exposition time.
Hunger dependent switching can contribute to the regulation and stabilization of the prey species preferred at a low hunger level.
The contribution of chemical control measures to the mortality of predacious mites can release populations of their prey species.
It can be advantageous to introduce, simultaneously with the predator, a harmless and rather unattractive alternative prey species.
Finally some suggestions are given to build a submodel of the multiple functional response to the densities of several kinds of prey into models for simulation of the dynamics of predator and prey populations.

Appendix I

Monte Carlo estimation of the mean and the standard deviation of random linear traverses through a unit square

A random linear traverse through a unit square may start at a point on one of the sides at an angle α with this side (Figure 32). The distance of this point from one of the corners of the square at the ends of the same side as the point, to be denoted by the length x, has a standard uniform probability distribution, with equal probabilities to values between zero and one. The angle α has a uniform probability distribution between zero and π radians. Depending on the values of x and α the length of the traverse L is determined by four different formulae. The different types of a traverse are indicated by $L_1 - L_4$ in Figure 32. The conditional relationships are:

$\alpha \leq 1/2\,\pi$ and $x\mathrm{tg}\alpha \leq 1 : L = x/\cos\alpha$
$\alpha \leq 1/2\,\pi$ and $x\mathrm{tg}\alpha \geq 1 : L = 1/\sin\alpha$
$\alpha \geq 1/2\,\pi$ and $(1-x)\mathrm{tg}(\pi-\alpha) \geq 1 : L = 1/\sin(\pi-\alpha)$
$\alpha \geq 1/2\,\pi$ and $(1-x)\mathrm{tg}(\pi-\alpha) \leq 1 : L = (1-x)/\cos(\pi-\alpha)$

The probability distribution of L, and hence its mean and standard deviation, are not easily derived from these relationships. A FORTRAN program was used, which generates random real numbers between zero and one to simulate random values of x and α (by multiplication with π). It uses the formulae given above to compute L. By this

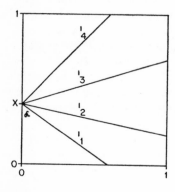

Fig. 32 | Linear traverses of a unit square.

program the mean and the standard deviation of five-thousand random linear traverses were computed. The output obtained was:

$\mu_L = 0.70887, \quad \sigma_L = 0.36374.$

Appendix II

A FORTRAN subroutine for the reclassing of fractions in compound simulation programs

For general application of compound simulation (see sections 4.3 and 4.4.3) the following subroutine may be used. The routine attaches a class number CLN to fractions, by comparing the fraction values of the values divided into classes with the class boundaries. Then the fractions are reclassed according to their class number.

The subroutine requires as input a matrix with the values of the state variables of the fractions (FVAR), an array with class boundaries (BO), the total number of fractions (NFT), an array with the numbers of classes of the variables divided into classes (NC), the number of variables divided into classes (NCV), and the total number of state variables considered (NVT). The output is a matrix with the values of the state variables of the classes (CVAR) after reclassing. The first row of the matrices FVAR and CVAR is allocated to the relative frequencies. The subscripted variables used are dimensioned to allow for nine state variables, three state variables divided into maximally five classes each, fifteen classes and thirty-five fractions. Of course the dimensions can be enlarged if this is desirable. We have:

```
SUBROUTINE RECLAS (FVAR,BO,NFT,
 NC,NCV,NVT,CVAR)
INTEGER CLN
DIMENSION FVAR(9,35),CVAR(9,15),
 BO(18),B(3,6),NC(3),CLN(35)
```

where B is an auxiliary matrix for the class boundaries.

The numbers of classes (NC) are redefined as unsuscripted integers, the total number of classes (NCT) is computed, and the class boundaries are rearranged in the matrix B, which is dimensioned for three variables with six class boundaries each. These introductory computations will facilitate the computation of the CLN values:

```
NCT=1
NC1=1
NC2=1
```

```
      NC3=1
      K=0
      DO 1 N=1,NCV
      IF (N.EQ.1) NC1=NC(N)
      IF (N.EQ.2) NC2=NC(N)
      IF (N.EQ.3) NC3=NC(N)
      NCT=NCT*NC(N)
      J=NC(N)+1
      DO 1 I=1,J
      K=K+1
    1 B(N,I)=BO(K)
```

Now the class numbers CLN can be determined for all fractions. In a
nest of DO loops the values of the variables of all fractions are compared with the class boundaries. If a class fits, its number is stored in
CLN and the program jumps out of the DO loops to continue with
the next fraction:

```
      DO 6 I=NFT
      IF (FVAR(1,I).EQ.0.0) GOTO 4
```

The program jumps to statement 4 if the relative frequency of the fraction I is zero.

```
      DO 3 J=1,NC1
      DO 3 K=1,NC2
      DO 3 L=1,NC3
      CLN(I)=J+NC1*(K-1)+NC1*NC2*(L-1)
```

This statement computes the class numbers.

```
      DO 2 N=1,NCV
      IF (N.EQ.1) M=J
      IF (N.EQ.2) M=K
      IF (N.EQ.3) M=L
    2 CLN(I)=CLN(I)*IOR(FVAR(N+1,I)-
      B(N,M),FVAR(N+1,I)-B(N,M+1))
      IF (CLN(I).GT.0) GOTO 6
    3 CONTINUE
    4 CLN(I)=1
    6 CONTINUE
```

IOR is a FORTRAN function which is one, if its first argument is positive or zero, and its second argument negative or zero. IOR is zero otherwise. The function is given below. The subroutine continues with reclassing and averaging, using the class numbers in the array CLN:

```
      DO 7 N=1,NCT
      DO 7 M=1,NVT
    7 CVAR(M,N)=0.0
      DO 8 I=1,NFT
      N=CLN(I)
      CVAR(1,N)=CVAR(1,N)+FVAR(1,I)
      DO 8 M=2,NVT
    8 CVAR(M,N)=CVAR(M,N)+
     FVAR(M,I)*FVAR(1,I)
      DO 10 N=1,NCT
      IF (CVAR(1,N).EQ.0.0) GOTO 10
      DO 9 M=2,NVT
    9 CVAR(M,N)=CVAR(M,N)/CVAR(1,N)
   10 CONTINUE
      RETURN
      END
```

The function IOR is compiled as a separate subprogram:

```
      FUNCTION IOR (X1,X2)
      IF (X1) 3,1,1
    1 IF (X2) 2,2,3
    2 IOR=1
      GOTO 4
    3 IOR=0
    4 RETURN
      END
```

Appendix III

A program for compound simulation of a predation process with prey eggs and males exposed simultaneously

```
TITLE       TYPRES/COMPOUND 1011712
PARAMETER   NE=80, NM=(1,2,4,6,8,10)
PARAMETER   DIGEST=0.4348
PARAMETER   IDIST=0.01, REPDEL=0.5
PARAMETER   CAETIM=0.012, CAMTIM=0.014
PARAMETER   MAXEC=0.94, MAXMC=0.67
CONSTANT    ZERO=1.E-05
FIXED       NCLASS,NFRACT,NCL,N,N1,N2,N3,
            M,NE,NM
STORAGE     NCL(3), BOUND(18)
TABLE       NCL(1-2)=5,3
TABLE       BOUND(1-6)=0.,.2,.4,.6,.8,1.08,
            BOUND(7-10)=-.5,.5,1.5,2.5
/           DIMENSION CLAV(9,15),FRAV(9,35)
MACRO       CATCHE,CATCHM=SEARCH(RELF,
            GUTCON,ACNE,ACNM,ACTMAL,...
            VELMAL,DENWEB)
```

Three decimal points at the end of the card indicate that the statement is continued on the next card.

```
            CATCHE=(1.0-EXP(-EC))*
             ECE/EC*RELF
            CATCHM=(1.0-EXP(-EC))*
             ECM/EC*RELF
            EC=AMAX1(1.E-05,ECE+ECM)
            ECE=ENCRAE*SUCRAE*ACTPRE*DELT
            ECM=ECA+ECRA
            ECA=(ENCRAA*SUCRAA+
             ENCRAR*SUCRAR)*ACTPRE*DELT
            ECRA=ENCRRA*SUCRRA*
             (1.0-ACTPRE)*DELT
            ENCRAE=COINAE*VELPRE*ACNE
            ENCRAA=COINAA*ACTMAL*SQRT
```

127

```
                   (VELPRE**2+VELMAL**2)*ACNM
              ENCRAR=COINAR*(1.0-ACTMAL)*
              VELPRE*ACNM
              ENCRRA=COINRA*ACTMAL*
              VELMAL*ACNM
              ACTPRE=ACTIM/(ACTIM+RESTIM)
```

The definitions of ACTIM, RESTIM, DISTUR, VELPRE, COINAA, COINAR, COINRA, COINAE, SUCRAA, SUCRAR, SUCRRA, and SUCRAE are given as sets of polynomials and orthonormal linear coefficients according to Table 9 in Section 3.9. They can be given in an arbitrary order, because the sorting procedure of the CSMP compiler will ensure, that they are computed in the right sequence.

```
ENDMAC
MACRO         ABAND,INGRT=EHAND(RELF,GUTCON,
              ACNM,RCTIM,EGG,ACTMAL,...
              DENWEB)
              IR=EIRT*REDHA*INSW(DELT-
              RCTIM,0.0,1.0-RCTIM/DELT)
              INGRT=INSW(EGG,0.,1.)*IR
              EIRT=79.2*(1.08-GUTCON)
```

The reduction factor REDHA is not a constant when males are present. Ignoring any influence on CAETIM we have:

```
              REDHA=EFTIM/(EHTIM-CAETIM)
              CAETIM=0.012
              ABAND=(1.0-EXP(-EA))*RELF
```

The relationship between EA, GUTCON, and IR is derived in Section 4.4.1 for the condition, that no prey males are present. If in this condition the mean feeding period and the mean handling period are EFTO and EHTO and the ingestion rate during handling periods is IRO, then:

$$IRO = \frac{EFTO \times (EHTIM-CAETIM) \times IR}{(EHTO-CAETIM) \times EFTIM}$$

EA will be reciprocally proportional to the handling time, since the latter will be as short as the abandonment probability is high:

128

$$EA = -\ln(1-0.013/(1.093-GUTCON))/0.054 \times IRO \times DELT \times$$
$$\times (EHTO-CAETIM)/(EHTIM-CAETIM)$$

This reduces to:

```
EA=-ALOG(1.-0.013/(1.093-
GUTCON))/0.054*IR*DELT*
EFTO/EFTIM
```

EHTIM, EFTIM, and EFTO can be defined according to Table 9.

```
ENDMAC
MACRO           ABAND,INGRT=MHAND(RELF,
                GUTCON,ACNM,RCTIM,MALE,
                DENWEB)
                IR=MIRT*REDHA*INSW(DELT-
                RCTIM,0.0,1.0-RCTIM/DELT)
                INGRT=INSW(MALE,0.,1.)*IR
                REDHA=MAFTIM/(MAHTIM-CAMTIM)
                CAMTIM=0.014
                MIRT=3.54*(1.08-GUTCON)
                ABAND=(1.0-EXP(-EA))*RELF
                EA=-ALOG(0.991-0.47*GUTCON**2-
                0.43*GUTCON)/0.054*IR*DELT
```

MAHTIM and MAFTIM are defined according to Table 9.

```
ENDMAC
MACRO           ACTMAL,VELMAL,DENWEB,DDIST=
                STATMA(ACNM,DIST)
                DENWEB=1.0-EXP(-0.028*DIST**
                (-0.13*ALOG(DIST)+0.88))
                DDIST=ACNM*VELMAL*ACTMAL*DELT
```

ACTMAL and VELMAL are defined according to Table 9

```
ENDMAC
INITIAL
NOSORT
                NCLASS=NCL(1)*NCL(2)
                NFRACT=NCL(1)*(NCL(2)+
                2*(NCL(2)-1))
                DO 2 N=1,NCLASS
                DO 1 M=1,9
```

```
      1        CLAV(M,N)=0.0
               CLAV(4,N)=NE
               CLAV(5,N)=NM
      2        CLAV(6,N)=IDIST
               CLAV(1,1)=1.0
DYNAMIC
NOSORT
               NPERT=0.0
               NPMRT=0.0
               BIOMRT=0.0
               MGUTRT=0.0
               ACNERT=0.0
               ACNMRT=0.0
               MDENW=0.0
               UTE=0.0
               UTM=0.0
               PROC=0.0
               REPL=IMPULS(0.0.REPDEL)
               DO 8 N=1,NCLASS
               N1=N+NCL(1)
               N2=N+NCL(1)*2
               N3=N+NCL(1)*4
               RELF=CLAV(1,N)
               IF (RELF.LT.ZERO) GOTO 6
               GUTCON=AMIN1(1.08,CLAV(2,N)*
                 INSW(CLAV(2,N),0.0,1.0))
               ACNE=CLAV(4,N)+REPL*
                 (NE-CLAV(4,N))
               ACNM=CLAV(5,N)+
                 REPL*(NM-CLAV(5,N))
               DIST=CLAV(6,N)
               RCTIM=CLAV(7,N)
               EGG=CLAV(8,N)
               MALE=CLAV(9,N)
               IF (N.GT.5) GOTO 3
SORT
               CATCHE,CATCHM=SEARCH(RELF,
                 GUTCON,ACNE,ACTMAL,...
                 VELMAL,DENWEB)
```

130

```
        ACTMAL,VELMAL,DENWEB,DDIST=
        STATMA(ACNM,DIST)
NOSORT
        FRAV(1,N)=RELF-CATCHE-CATCHM
        FRAV(1,N1)=CATCHE
        FRAV(1,N2)=CATCHM
        FRAV(2,N)=GUTCON-DIGEST*
        GUTCON*DELT
        FRAV(2,N1)=FRAV(2,N)
        FRAV(2,N2)=FRAV(2,N)
        FRAV(3,N)=0.0
        FRAV(3,N1)=1.0
        FRAV(3,N2=2.0
        FRAV(4,N)=ACNE
        FRAV(4,N1)=AMAX1(0.0,ACNE-1.0)
        FRAV(4,N2)=ACNE
        FRAV(5,N)=ACNM
        FRAV(5,N1)=ACNM
        FRAV(5,N2)=AMAX1(0.0,ACNM-1.0)
        FRAV(6,N)=DIST+DDIST
        FRAV(6,N1)=FRAV(6,N)
        FRAV(6,N2)=FRAV(6,N)
        FRAV(7,N)=0.0
        FRAV(7,N1)=CAETIM
        FRAV(7,N2)=CAMTIM
        FRAV(8,N)=0.0
        FRAV(8,N1)=MAXEC
        FRAV(8,N2)=0.0
        FRAV(9,N)=0.0
        FRAV(9,N1)=0.0
        FRAV(9,N2)=MAXMC
        NPERT=NPERT+CATCHE/DELT
        NPMRT=NPMRT+CATCHM/DELT
        MGUTRT=MGUTRT-DIGEST*GUTCON*
        RELF
        ACNERT=ACNERT-CATCHE/DELT
        ACNMRT=ACNMRT-CATCHM/DELT
        MDENW=MDENW+RELF*DENWEB
        GOTO 8
```

```
      3         IF (N. GT.10) GOTO 4
SORT
                ABAND,INGRT=EHAND(RELF,GUTCON,
                 ACNM,ACTIM,EGG,ACTMAL,...
                 DENWEB)
                ACTMAL,VELMAL,DENWEB,
                 DDIST=STATMA(ACNM,DIST)
NOSORT
                FRAV(3,N2)=1.0
                UTE=UTE+RELF*INGRT/(ACNE+1.0)
                GOTO 5
      4         CONTINUE
SORT
                ABAND,INGRT=MHAND(RELF,GUTCON,
                 ACNM,RCTIM,MALE,DENWEB)
                ACTMAL,VELMAL,DENWEB,
                 DDIST=STATMA(ACNM,DIST)
NOSORT
                FRAV(3,N2)=2.0
                UTM=UTM+RELF*INGRT/(ACNM+1.0)
      5         FRAV(1,N2)=RELF-ABAND
                FRAV(1,N3)=ABAND
                FRAV(2,N2)=GUTCON-DIGEST*
                 GUTCON*DELT+INGRT*DELT
                FRAV(2,N3)=FRAV(2,N2)
                FRAV(3,N3)=0.0
                FRAV(4,N2)=ACNE
                FRAV(4,N3)=ACNE
                FRAV(5,N2)=ACNM
                FRAV(5,N3)=ACNM
                FRAV(6,N2)=DIST+DDIST
                FRAV(6,N3)=FRAV(6,N2)
                FRAV(7,N2)=RCTIM-INSW
                 (DELT-RCTIM,DELT,RCTIM)
                FRAV(7,N3)=0.0
                FRAV(8,N2)=AMAX1(0.0,EGG-
                 INGRT*DELT)
                FRAV(8,N3)=0.0
```

```
              FRAV(9,N2)=AMAX1(0.0,MALE-
                INGRT*DELT)
              FRAV(9,N3)=0.0
              BIOMRT=BIOMRT+RELF*INGRT
              MGUTRT=MGUTRT-
                DIGEST*GUTCON*RELF+RELF*INGRT
              MDENW=MDENW+RELF*DENWEB
              PROC=PROC+RELF/DELT
              GOTO 8
      6       IF (N.GT.5) GOTO 7
              FRAV(1,N)=0.0
              FRAV(1,N1)=0.0
              FRAV(1,N2)=0.0
              GOTO 8
      7       FRAV(1,N2)=0.0
              FRAV(1,N3)=0.0
      8       CONTINUE
              CALL RECLAS(FRAV,BOUND,
                NFRACT,NCL,2,9,CLAV)
              NPE=INTGRL(0.0,NPERT)
              NPM=INTGRL(0.0,NPMRT)
              BIOMC=INTGRL(0.0,BIOMRT)
              MGUTC=INTGRL(0.0,MGUTRT)
              MACNE=INTGRL(NE,ACNERT+
                REPL*(NE-MACNE)/DELT)
              MACNM=INTGRL(NM,ACNMRT+
                REPL*(NM-MACNM)/DELT)
              IMORTE=NPERT/AMAX1(ZERO,
                MACNE)*5.0
              IMORTM=NPMRT/AMAX1(ZERO,
                MACNM)*5.0
TIMER         DELT=0.01, PRDEL=1,0,
                FINTIM=24.0
METHOD        RECT
PRINT         NPE,NPM,MGUTC,BIOMC,MACNE,
              MACNM,IMORTE,IMORTM,MDENW,...
              UTE,UTM,PROC
END
STOP
```

Appendix IV

Names of variables and their dimensions

A B A N D	Indicator or relative frequency of abandonment of prey in D E L T
A C N	Actual number of prey present on a leaf disk of five cm^2
A C T I M (h)	Mean period of walking of the predator
A C T M A L	Activity of the males, the proportion of males walking
A C T P R E	Activity of the predator, the proportion of time spent walking
B I O M C (egg eq.)	Biomass consumed
B I O M R T (egg eq. h^{-1})	Rate of biomass consumption
B O U N D	Matrix of the boundaries of the classes in the compound simulation models
C A T C H	Indicator or relative frequency of a capture in D E L T
C A T I M (h)	Time period needed by the predator to catch a prey
C L A V	Matrix of the variable values of the classes in the compound simulation models
C O I N (m^{-1})	Coincidence in space of the predator and a prey
D D I S T (m)	Increment of the total distance covered by prey males in D E L T
D E L T (h)	Time-interval used for simulation
D E N E G G (cm^{-2})	Density of the prey eggs
D E N M A L (cm^{-2})	Density of the prey males
D E N W E B	Relative density of the webbing cover
D I G E S T (h^{-1})	Digestion rate per egg equivalent of the gut content
D I G R T (egg eq. h^{-1})	Digestion rate
D I S T (m)	Total distance covered by prey males on the leaf disk of five cm^2

DISTUR	Proportion of encounters of a resting predator causing reactivation by disturbance
EA	Expectation value of the number of abandonments in DELT
EC	Expectation value of the number of captures in DELT
EGG (egg eq.)	Actual amount of food available in a prey egg
ENCRT (h^{-1})	Encountering rate
FINTIM (h)	The limit of the sum of time-steps, at which time the simulation is terminated
FRAV	Matrix of the variable values of fractions in the compound simulation models
GUTCON (egg eq.)	The gut content of the predator
H	Indicator of the engagement of the predator, which is one when this is handling prey and zero otherwise
HTIM (h)	Period of handling prey
IMORT ($cm^2 h^{-1}$)	Intrinsic rate of mortality of the prey population, which is the rate of increase of the area cleared of prey by one predator
INGEST (h^{-1})	Ingestion rate per egg equivalent of the gut capacity
INGRT (egg eq. h^{-1})	Ingestion rate
IR (egg eq. h^{-1})	Ingestion rate when the predator handles a prey which is not empty
IRT (egg eq. h^{-1})	Ingestion rate when the predator actually feeds on a prey
MACN	Expectation value of ACN
MALE (egg eq.)	Actual amount of food available in a prey male
MAXC (egg eq.)	Maximum food content of a prey
MAXGUT (egg eq.)	Maximum gut content
MDENW	Expectation value of DENWEB
MGUTC (egg eq.)	Expectation value of GUTCON
N	Number of prey maintained on a leaf disk of five cm^2
NCLASS	Total number of classes in the compound simulation models
NFRACT	Total number of fractions in the compound simulation models

NP	Number of prey captured
PR (h^{-1})	Predation rate, which is the number of prey captured per time unit
PRA	Probability of an abandonment in DELT
PRC	Probability of a capture in DELT
PRDEL (h)	Time interval for the printing of output during simulation
PREDRT (h^{-1})	Predation rate, the number of prey captured per hour
PROC	Proportion of time the predator is occupied with handling prey
PSPEC	Prey species
RCTIM (h)	Remnant of the time period needed to catch a prey
RCTRT	Rate of change of RCTIM
REDHA	Reduction factor of the ingestion rate, accounting for other handling components than feeding
RELF	Relative frequency
REPDEL (h)	Time interval for prey replenishment
REPL	Number of prey replenished in DELT
RESTIM (h)	Mean period of resting of the predator
RN	Random number between zero and one
S	Indicator of the engagement of the predator, which is one when the predator is searching for prey and zero otherwise
SUCRT	Success ratio
UT (cm^2 h^{-1})	The utility of the prey population, which is the rate of increase of the area cleared of prey biomass by one predator
VELMAL (m h^{-1})	The locomotion velocity of the prey males
VELPRE (m h^{-1})	The locomotion velocity of the predator

Table of prefixes and suffixes

−AA	Active predator and active prey male
−AR	Active predator and resting prey male
−AE	Active predator and prey egg
−E−	Prey egg

136

−P	Polinomial
−F	Orthonormal factor
−M, MA−	Prey male
−RA	Resting predator and active prey male

Acknowledgments

I wish to express my gratitude to Professor H. Klomp and Professor C. T. de Wit for the opportunity to study the causal-analytical approach in the investigation of ecological processes, for their stimulating guidance of the work, and for the critical reading of the manuscript. I am also greatly indepted to Drs J. H. Kuchlein for his co-operation, and to all the workers and students at the Laboratory of Zoology, the Laboratory of Phytopathology and the Department of Theoretical Production Ecology of the Agricultural University and at the Institute of Phytopathological Research at Wageningen, who contributed to this study.

References

Bailey, N. T. J., 1967. The mathematical approach to biology and medicine. Wiley, New York.

Ballard, R. C., 1954. The biology of the predacious mite *Typhlodromus fallacis* (Garman) (Phytoseiidae) at 78 °F. Ohio J. Sci. 54: 175–179.

Bartlett, M. S., 1960. Stochastic population models in ecology and epidemiology. Methuen & Co., London.

Beukema, J. J., 1968. Predation by the three-spined stickleback (*Gasterosteus aculeatus* L.): the influence of hunger and experience. Behaviour 31: 1–126.

Beverton, R. J. & S. J. Holt, 1957. On the dynamics of exploited fish populations. Fishery Invest., Lond. II, 19: 1–533.

Bravenboer, L., 1959. The chemical and biological control of the spider mite *Tetranychus urticae* Koch. Diss. Agric. Univ. Wageningen.

Bravenboer, L. & G. Dosse, 1962. *Phytoseiulus riegeli* Dosse as a predator of harmful mites of the group of *Tetranychus urticae*. Entomologia exp. appl. 5: 291–304.

Brennan, R. D., C. T. de Wit, W. A. Williams & E. V. Quattrin, 1970. The utility of a digital simulation language for ecological modelling. Oecologia 4: 113–132.

Burnett, T., 1970. Effect of temperature on a greenhouse acarine predator-prey population. Can. J. Zool. 48: 555–562.

Chant, D. A., 1959. Phytoseiid mites (Acarina). I. Bionomics of seven species in southeastern England. II. A taxonomic review of the family Phytoseiida, with descriptions of 38 new species. Can. Ent. 91; suppl. 12.

Chant, D. A., 1961a. An experiment in biological control of *Tetranychus telarius* (L.) (Acarina: Tetranychidae) in a greenhouse using the predacious mite *Phytoseiulus persimilis* Athias-Henriot (Phytoseiidae). Can. Ent. 93: 437–443.

Chant, D. A., 1961b. The effect of prey density on prey consumption and oviposition in adults of *Typhlodromus* (T.) *occidentalis* Nesbitt (Acarina: Phytoseiidae) in the laboratory. Can. J. Zool. 39: 311–315.

Chant, D. A., 1963. Some mortality factors and the dynamics of orchard mites. Mem. ent. Soc. Can. 32.

Chiang, C. L., 1968. Introduction to stochastic processes in biostatistics. Wiley, New York.

Collyer, E., 1958. Some insectary experiments with predacious mites to determine their effect on the development of *Metatetranychus ulmi* (Koch) populations. Entomologia exp. appl. 1: 138–146.

Dixon, A. F. G., 1959. An experimental study of the searching behaviour of the predatory coccinellid beetle *Adalia decempunctata* (L.). J. Anim. Ecol. 28: 259–281.

Dosse, G., 1956. Über die Entwicklung einiger Raubmilben bei verschiedene Nahrungstieren (Acar., Phytoseiidae). Pflanzenschutzberichte 16: 122–136.

Elbadry, E. A., A. M. Afifi, G. I. Issa & E. M. Elbenhawy, 1968. Effect of different prey species on the development and fecundity of the predacious mite, *Amblyseius gossipi* (Acarina: Phytoseiidae). Z. angew. Ent. 62: 247–252.

Feller, W., 1968. An introduction to probability theory and its application, Vol. I, 3d edn. Wiley, New York.

Ferrari, Th. J., 1952. An agronomic research with potatoes on the river ridge soils of the Bommelerwaard. Diss. Agric. Univ. Wageningen.

Flaherty, D. L. & C. B. Huffaker, 1970a. Biological control of pacific mites and willamette mites in San Joaquin Valley vineyards. I. Role of *Metaseiulus occidentalis*. Hilgardia 40: 267–308.

Flaherty, D. L. & C. B. Huffaker, 1970b. Biological control of pacific mites and willamette mites in San Joaquin Valley vineyards. II. Influence of dispersion patterns of *Metaseiulus occidentalis*. Hilgardia 40: 309–330.

Fleschner, C. A., 1950. Studies on searching capacity of the larvae of three predators of the citrus red mite. Hilgardia 20: 233–265.

Foot, W. H., 1963. Competition between two species of mites. II. Factors influencing intensity. Can. Ent. 95: 45–57.

Garfinkel, D. A., 1965. Computer simulation in biochemistry and ecology. In: Theoretical and mathematical biology. T. H. Waterman & H. J. Morowitz (eds.), Blaisdell Publ. Comp., New York, p. 292–310.

Gasser, R., 1951. Zur Kenntnis der gemeinen Spinnmilbe *Tetranychus urticae* Koch. Bull. Soc. ent. Suisse 24: 217–262.

Gervet, J., 1968. Interaction entre individus et phénomène social. Archs néerl. Zool. 18: 205–252.

Haynes, D. L. & P. Sisojevic, 1966. Predatory behaviour of *Philodromus rufus* Walckenaer (Aranea, Thomisidae). Can. Ent. 98: 113–133.

Herbert, H. J., 1956. Laboratory studies on some factors in the life history of the predacious mite *Typhlodromus tiliae* Oudms. (Acarina: Phytoseiidae). Can. Ent. 88: 701–704.

Hinde, R. A., 1954. Changes in responsiveness to a constant stimulus. Br. J. Anim. Behav. 2: 41–55.

Holling, C. S., 1959a. The components of predation as revealed by a study of small-mammal predation of the European pine sawfly. Can. Ent. 91: 293–320.

Holling, C. S., 1959b. Some characteristics of simple types of predation and parasitism. Can. Ent. 91: 385–398.

Holling, C. S., 1961. Principles of insect predation. A. Rev. Ent. 6: 163–182.

Holling, C. S., 1963. An experimental component analysis of population processes. Mem. ent. Soc. Can. 32: 22–32.

Holling, C. S., 1964. The analysis of complex population processes. Can. Ent. 96: 335–347.

Holling, C. S., 1965. The functional response of predators to prey density and its role in mimicry and population regulation. Mem. ent. Soc. Can. 45: 60 pp.

Holling, C. S., 1966. The functional response of invertebrate predators to prey density. Mem. ent. Soc. Can. 48: 86 pp.

Huffaker, C. B., M. van de Vrie & J. A. McMurtry, 1969. The ecology of tetranychid mites and their natural control. A. Rev. Ent. 14: 125–174.

Huffaker, C. B., M. van de Vrie & J. A. McMurtry, 1970. Ecology of tetranychid mites and their natural enemies: a review. II. Tetranychid populations and their possible control by predators: an evaluation. Hilgardia 40: 391–458.

IBM, 1968. System/360 Continuous System Modeling Program, user's manual.

Kowal, N. E., 1971. A rationale for modeling dynamic ecological systems. In: Systems analysis and simulation in ecology, Vol. I, B. C. Patten (ed.), Acad. Press, New York, p. 123–194.

Kuchlein, J. H., 1966. Mutual interference among the predacious mites *Typhlodromus longipilus* Nesbitt (Acari: Phytoseiidae). I. Effects of predator density on oviposition rate and migration tendency. Med. Landb. Wet. Gent 31: 740–746.

Laing, J. E., 1969a. Life history and life table of *Metaseiulus occidentalis*. Ann. ent. Soc. Am. 62: 978–982.

Laing, J. E., 1969b. Life history and life table of *Tetranychus urticae* Koch. Acarology 11: 32–42.

141

Laing, J. E. & C. B. Huffaker, 1969. Comparative studies of predation by *Phytoseiulus persimilis* Athias-Henriot and *Metaseiulus occidentalis* (Nesbitt) (Acarina: Phytoseiidae) on populations of *Tetranychus urticae* Koch (Acarina: Tetranychidae). Res. Popul. Ecol. 11: 105–126.

Lee, M. G. & D. W. Davis, 1968. Life history and behaviour of the predatory mite *Typhlodromus occidentalis* in Utah. Ann. ent. Soc. Am. 61: 251–255.

McMurtry, J. A., C. B. Huffaker & M. van de Vrie, 1970. Ecology of Tetranychid mites and their natural enemies, a review. I. Tetranychid enemies: their biological characters and the impact of spray practices. Hilgardia 40: 331–390.

Mori, H., 1969. The influence of prey-density on the predation of *Amblyseius longispinosus* (Evans) (Acarina: Phytoseiidae) Proc. 2nd Int. Congr. Acarology, Sutton-Bonington 1967, p. 149–153.

Mori, H. & D. A. Chant, 1966. The influence of prey density, relative humidity, and starvation on the predacious behaviour of *Phytoseiulus persimilis* Athias-Henriot (Acarina: Phytoseiidae). Can. J. Zool. 44: 483–491.

Mukerji, M. K. & E. J. LeRoux, 1969. A quantitative study of food consumption and growth of *Podisus maculiventris* (Hemiptera: Pentatomidae). Can. Ent. 101: 387–403.

Murdoch, W. W., 1969. Switching in general predators: experiments on predator specificity and stability of prey populations. Ecol. Monogr. 39: 335–354.

Murdoch, W. W., 1973. The functional response of predators. J. appl. Ecol. 11: 335–342.

Nakamura, K., 1968. The ingestion in wolf spiders. I. Capacity of gut of *Lycosa pseudoannulata*. Res. Popul. Ecol. 10: 45–53.

Nicholson, A. J., 1933. The balance of animal populations. J. Anim. Ecol. 2: 132–178.

Patten, B. C., 1971. A primer for ecological modeling and simulation with analog and digital computers. In: *Systems analysis and simulation in ecology*, Vol. I, B. C. Patten (ed.), Acad. Press, New York, p. 3–121.

Putman, W. L., 1962. Life-history and behaviour of the predacious mite *Typhlodromus* (T.) *caudiglans* Schuster (Acarina: Phytoseiidae) in Ontario, with notes on the prey of related species. Can. Ent. 94: 163–177.

Rashevsky, N., 1959. Some remarks on the mathematical theory of nutrition of fishes. Bull. math. Biophys. 21: 161–182.

142

Rivard, I., 1962. Some effects of prey density on survival, speed of development, and fecundity of the predaceous mite *Melichares dentriticus* (Berl.) (Acarina: Aceosejidae). Can. J. Zool. 40: 1233–1236.

Royama, T., 1971. A comparative study of models for predation and parasitism. Res. Popul. Ecol. suppl. 1.

Ruiter, L. de, 1956. The measurement of the prey-value of preys. Archs néerl. Zool. 11: 524–526.

Sandness, J. N. & J. A. McMurtry, 1970. Functional response of three species of Phytoseiidae (Acarina) to prey density. Can. Ent. 102: 692–704.

Sandness, J. N. & J. A. McMurtry, 1972. Prey consumption of *Amblyseius largoensis* in relation to hunger. Can. Ent. 104:461–470.

Skellam, J. G., 1958. The mathematical foundations underlying the use of line transects in animal ecology. Biometrics 14:385–400.

Smith, J. C. & L. D. Newson, 1970. Laboratory evaluation of *Amblyseius fallacis* as a predator of Tetranychid mites. J. econ. Ent 63: 1876–1878.

Solomon, M. E., 1949. The natural control of animal populations. J. Animal Ecol. 18: 1–35.

Stigter, C. J., 1973. Physics and practical use of some instruments and methods in agricultural meteorology. Lecture notes M.sc. course on soil science and water management, Agricultural University, Wageningen.

Thorpe, W. H., 1962. Learning and instinct in animals. Methuen, London.

Tinbergen, L. & H. Klomp, 1960. The natural control of insects in pine woods. II. Conditions for damping of Nicholson oscillations in parasite-host systems. Archs néerl. Zool. 13:344–379.

Tocher, K. D., 1963. The art of simulation. English Univ. Press, London.

Voûte, A. D., 1946a. Regulation of the density of the insect-populations in virgin forests and cultivated woods. Archs néerl. Zool. 7: 435–470.

Voûte, A. D., 1946b. Toegepaste entomologie en verrijking of verarming der fauna. Tijdschr. PlZiekt. 2: 150–157.

Watt, K. E. F., 1961. Mathematical models for use in insect pest control. Can. Ent. 93: suppl. 19.

Watt, K. E. F., 1966. The nature of systems analysis. In: Systems analysis in ecology, K. E. F. Watt (ed.), Acad. Press, New York, p. 1–14.

Watt, K. E. F., 1968. Ecology and resource management, a quantitative approach. McGraw-Hill, New York.

Wolda, H., 1961. Response decrement in the prey catching activity of *Notonecta glauca* L. (Hemiptera). Archs néerl. Zool. 14: 61–89.

Wynne-Edwards, V. C., 1962. Animal dispersion in relation to social behaviour. Oliver & Boyd, Edinburgh.